Rapid Load Testing on Piles

T0173999

Rapid Load Testing on Piles

Editors

Paul Hölscher

Deltares, Delft, The Netherlands

Frits van Tol

Delft University of Technology, Delft, The Netherlands
Department of Civil Engineering and Geosciences, Deltares, Delft, The Netherlands

CRC Press
Taylor & Francis Group
Boca Raton London New York

CRC Press is an imprint of the
Taylor & Francis Group, an **informa** business

A BALKEMA BOOK

CRC Press
Taylor & Francis Group
6000 Broken Sound Parkway NW, Suite 300
Boca Raton, FL 33487-2742

First issued in paperback 2019

© 2009 by Taylor & Francis Group, LLC
CRC Press is an imprint of Taylor & Francis Group, an Informa business

No claim to original U.S. Government works

ISBN-13: 978-0-415-48297-4 (hbk)
ISBN-13: 978-0-367-38629-0 (pbk)

This book contains information obtained from authentic and highly regarded sources. Reasonable efforts have been made to publish reliable data and information, but the author and publisher cannot assume responsibility for the validity of all materials or the consequences of their use. The authors and publishers have attempted to trace the copyright holders of all material reproduced in this publication and apologize to copyright holders if permission to publish in this form has not been obtained. If any copyright material has not been acknowledged please write and let us know so we may rectify in any future reprint.

Except as permitted under U.S. Copyright Law, no part of this book may be reprinted, reproduced, transmitted, or utilized in any form by any electronic, mechanical, or other means, now known or hereafter invented, including photocopying, microfilming, and recording, or in any information storage or retrieval system, without written permission from the publishers.

For permission to photocopy or use material electronically from this work, please access www.copyright.com (http://www.copyright.com/) or contact the Copyright Clearance Center, Inc. (CCC), 222 Rosewood Drive, Danvers, MA 01923, 978-750-8400. CCC is a not-for-profit organization that provides licenses and registration for a variety of users. For organizations that have been granted a photocopy license by the CCC, a separate system of payment has been arranged.

Trademark Notice: Product or corporate names may be trademarks or registered trademarks, and are used only for identification and explanation without intent to infringe.

Typeset by Vikatan Publishing Solutions (P) Ltd, Chennai, India.

Published by: CRC Press/Balkema
P.O. Box 447, 2300 AK Leiden, The Netherlands
e-mail: Pub.NL@taylorandfrancis.com
www.crcpress.com – www.taylorandfrancis.co.uk – www.balkema.nl

Visit the Taylor & Francis Web site at
http://www.taylorandfrancis.com

and the CRC Press Web site at
http://www.crcpress.com

Contents

Delft Cluster project

Preface

The bearing capacity of a foundation pile depends almost solely on the soil in which the pile is installed. Although the foundation engineer strongly relies on the correlation between the bearing capacity of a pile and the results of field-tests (such as CPT and SPT), there has been strong desire for more empirical information on the real bearing capacity of foundation piles. Many contractors develop new cast-in-place systems, progress in this field would also benefit from more intensive testing of the actual bearing capacity.

Unfortunately, static load tests on piles are time consuming and expensive. The limited number of tests hinders the development of cheaper and more reliable foundation systems, the delivery of higher added value to the market by contractors and an adequate quality control by principals and building authorities. A rapid load test may form a good alternative for a static load test; it is faster and therefore economically more attractive.

Before rapid load testing on piles can be embraced however, there is a need for validation of this method to ensure that it forms a decent alternative for a static load test. Moreover, there is a need for a standard prescribing a consequent test execution, and a guideline enforcing a proper interpretation of test results; all together, a uniform, international test standard should be developed.

We started our research at Delft University of Technology and GeoDelft (now part of Deltares) as a Dutch project, following a grant of Delft Cluster and a grant of the World Bank for a PhD-student. Considering the importance of international guidelines in today's globalizing world and the fact that regulation should meet the European Euro codes, it soon became a joint project of European researchers. We organized two seminars, to discuss the progress in rapid load testing and the need for regulation. These seminars, originally intended for Europe only, were also attended by researchers from Japan and the USA. Following these meetings, all partners agreed to write a contribution presenting an overview of their research, the core results and the implications for the engineering practice.

This book contains the expert contributions by worldwide leading researchers. The effects of various factors are investigated, such as the influence of the loading rate, pore water pressure and the test's reliability under field conditions. We have authored general information on rapid load testing and results of field tests and centrifuge testing to complete the contents. The book presents an accurate description of the execution of the rapid load test.

Paul Hölscher
Frits van Tol

Acknowledgments

The financial support of the following partners is kindly acknowledged:

- Ministry of Transport, Public Works and Water Management
- IHC Hydrohammer
- Shell Global Solutions
- VWS Geotechniek
- Ballast Nedam

The grant of the Delft Cluster for this project is kindly acknowledged.

All partners in the project, both national as international, who supported the project with their activities, are kindly acknowledged. Especially all authors who wrote a paper for the seminar, played an essential role in the realization of this book.

The financial support of Berminghammer Inc. for the creation of this book is kindly acknowledged.

Rapid Load Testing on Piles

P. Hölscher
Deltares, Delft, The Netherlands

A.F. van Tol
Delft University of Technology, Delft, The Netherlands
Department of Civil Engineering and Geosciences, Deltares, Delft, The Netherlands

SUMMARY

This paper will give an overview of the state of the art on rapid load testing and the background of the second seminar on rapid load testing in Delft. The history of the test is shortly presented. The benefits and differences of the rapid load test are compared with the Static load test and dynamic load test.

The main scientific questions: the rate dependency of clay, excess pore water pressures in sand and the interpretation models, are shortly discussed. The standard for execution of rapid load tests and guideline for interpretation are presented.

This introduction includes a short introduction of the advanced literature in this book.

1 INTRODUCTION

1.1 Background of the research

The rapid load test seems to be a good and economical alternative for the Static load test on piles. Examples of this test method are Statnamic (Janes *et al.* 1991) and the Pseudo-Static Pile Load Tester (Schellingerhout & Revoort 1996). However, in Europe the application of such tests is hindered by the discussion about the interpretation of the test. This hindrance can be overcome by proper regulation of the test, and an international project has been started with the objective of developing a Standard for the execution of a rapid load test and Guidelines for the interpretation of the test (Hölscher 2007).

1.2 Short history of the test

The first rapid load test on piles was carried out by [Gonin *et al.* 1984]. The impact force is developed by dropping a heavy mass. The testing method is named Dynatest.

[Birmingham & Janes 1989] developed an innovative loading system, which loads the pile by launching a reaction mass. The testing method is named as Statnamic. The mass is launched by burning rocket fuel in a closed burning chamber. The inertia of the reaction mass creates the force on the pile head. After some time, the force decreases quickly due to the increase volume of the burning chamber (the mass moves up) and the exhaustion of the fuel. Due to the properties of the system (burning rocket fuel), always a smooth increase and decrease of the force is obtained.

Before starting the test, the reaction mass is supported by the pile head. Once the fuel is burned and exhausted, the reaction mass falls down. It must be prevented that the reaction mass hits the pile head after the test. Therefore, a catching mechanism is developed. The older systems use a catching mechanism with gravel. Recently, a mechanical catching mechanism has been developed for systems up to 4 MN. This mechanism is also used to place the reaction mass on the pile, offering the possibility to do a number of tests on one pile. Nowadays, systems up to 16 MN pile load are developed; Statnamic is the system with the highest possible working load.

[Schellingerhout *et al.* 1996] developed the Pseudo-Static Pile Load Tester (PSPLT). In this equipment, a special set of springs is mounted to the falling mass. This set of springs is non-linear. The non-linearity is essential to create a smooth loading and unloading of the pile head. After the mass has bounced on the pile head, it is caught by the PSPLT for preventing rebounce.

The complete system (mass with springs) is mounted in a frame and efficiently transported by machinery. This facilitates a high number of tested piles per day, and offers the possibility to test one pile with several drop heights. For higher peak loads, a new set of springs has to be developed.

The PSPLT is developed for on-shore applications. The full machinery is very flexible at a building site. The Statnamic method is efficient for offshore applications, since the reaction mass is supported by the pile and relatively high forces can be reached. The maximum working load for Statnamic equipment is higher.

A method for smaller piles with the cushion on the pile head, which is normal for dynamic pile testing, is under development in Japan (Matsumoto 2008).

1.3   Principle of the test

The essential principle of the rapid load test on a pile is the duration of the dynamic loading on the pile head, relative to the duration that a stress wave needs to travel from the pile head to the pile toe. A dynamic load is considered a rapid load if

–   the loading is first continuously increasing and then continuously decreasing in a sufficiently smooth manner.
–   the load duration should be that long that the load does not induce a wave in the pile. The duration of loading should be significantly longer than a wave needs to travel from the pile head to the pile toe.

If the duration of the load is shorter than the prescribed value, the test cannot be considered a rapid load test and should be treated as a dynamic load test (DLT).

Due to the long duration of the loading, the risk of tension in the pile is minimal. This is important for testing cast-in-place piles, which normally cannot withstand tension forces.

The relatively high force can be generated by using a relatively small mass, by accelerating it during a sufficient short time interval. The shorter the time for reaching a given velocity, the higher the acceleration of the mass, the higher the force.

Table 1 Characteristics of pile tests.

Type of test	Static	Rapid	Dynamic
duration of load	16 hrs	100 ms	7 ms
number of test per day	1	2	8
reaction mass needed (perc. bearing capacity)	100%	5–10%	2%
time needed for result	directly	10 min.	4 hours
tension-stress in pile	no	no	possible
prefab pile	yes	yes	yes
cast-in-place	yes	yes	no
stress in soil	static	dynamic	dynamic
pore water pressure in sand	absent	occurs	occurs
costs (Euro per ton)	100	20	8
reliability	high	unknown	reasonable

At this moment, two principles are used for rapid load testing:

– the falling mass, as used originally by Gonin and Revoort. In Japan, new, cheap methods are under development using this principle. The choice of a proper spring is essential for this method. Creation of a sufficient smooth loading is complex.
– the launched mass method, as developed by Berminghammer. This method is known under the name Statnamic. The fuel generates more or less automatically a smooth loading at the pile head.

The aim of a load test is to proof that a foundation fulfills the specified requirements. Normally, these requirements are specified in terms of the stiffness under working load and the bearing capacity in ultimate limit state. Principals are interested in these data, since they want to be sure that the installed foundation is adequate, whereas contractors are interested in showing that their product fulfills the contractual requirements.

In general three types of tests are available:

– static load test
– rapid load test
– dynamic load test

Table 1 summarizes the main characteristics of these three tests.

The rapid load test seems to give a reasonable quality for its price. However, acceptance is low in Europe. The research described in this paper tries to improve the acceptance of the test for practical engineering.

2   WHY MORE ECONOMIC TEST METHODS?

The application of pile foundations in delta areas with soft soil is a necessary condition to be able to construct stable and durable structures. In the Netherlands, the bearing capacity of piles is calculated using empirical formulae using cone penetration data. This procedure is described in the Dutch codes. In most European countries, many static pile tests are carried out.

Several actual developments lead to the necessity to reconsider this way of doing.

– Increase of the number of alternative piling systems and the increase of the free market system. More and more, contractors are trying to develop and use cheaper and better systems. The performance and bearing capacity of such piles strongly depends on workmanship of the contractor, leading to an increasing demand on possibilities to check the piles by the building authorities.
– Increase of the complexity of (foundation) structures. The societal requirements lead to more complex structures, which are far beyond the practical experience. Uncertainties in design models lead to designs that are too robust and therefore too costly, or designs that have much too high risk factors. The design can be optimized, but only if more knowledge about performance is available and complex foundations can be tested. This requires a reasonable priced and generally accepted method for measuring performance and bearing capacity.
– Re-usage of existing piles. Often, existing structures and buildings are replaced by new ones. Normally, the existing foundation is demolished and rebuilt, since no reliable data on the bearing capacity of the existing foundation is available. In these cases, a gain of money and execution time can be reached if the existing foundation can be re-used after measuring its bearing capacity.
– For tension piles and horizontally loaded piles actually no good testing methods for practical application are available.
– The introduction of Eurocode 7 is an excellent moment for introducing a quick and cheap testing method for bearing capacity and reliability of foundation piles.

In the Netherlands, two more developments must be noted

– Introduction of Eurocode 7.
– Increase of the application of the cast-in-place piles.

The influence of other European countries is noted: the results of field-tests are more important than in the Dutch codes. The cone penetration test is not considered anymore as the ultimate geotechnical test. The Eurocode 7 gives a clear description when piles shall be tested:

– When using a type of pile or installation method for which there is no comparable experience;
– When the piles have not been tested under comparable soil and loading conditions;
– When the piles will be subjected to loading for which theory and experience do not provide sufficient confidence in design;
– When observations during the process of installation indicates pile behavior that deviates strongly and unfavorable from the behavior anticipated based on the site investigation or experience, and when additional ground investigations do not clarify the reasons for this deviation.

The amount of construction activity in populated areas increases. The acceptance of vibrations from pile driving decreases. Therefore, more and more piling systems with low vibrations are applied. This leads to an increase in the application of cast-in-place piles. The quality of these systems strongly depends on workmanship, which leads to a demand for cheap methods for quality check of an installed pile.

Better knowledge of pile bearing capacity offers the possibility of an economic pile design. Consequently, the difference between the pile failure and soil failure will be smaller and the concrete in the pile will be loaded to a higher level. For pile testing in general, the linearity of pile behavior must be evaluated carefully. In particular for rapid testing and dynamic testing, since the concrete stresses in the pile are much higher than during static testing. This is discussed in detail by (Stokes and Mullins 2008).

3   DESCRIPTION OF THE DELFT RESEARCH PROJECT

The question why the rapid load test is not used in the Netherlands was posed to a group of experts in the field of foundation engineering. Their answers gave the following important reasons for some distrust in the Netherlands:

– no calibrated interpretation method is available
– the influence of the pore water pressure on bearing capacity is not known
– the influence of rate effects are not known
– no evidence of reliability is available.

Another disadvantage, which was mentioned, is the lack of gain by doing field tests in current regulations.

The Delft Cluster research was started, focusing on answering these questions.

3.1   Scientific questions

A literature review to the rate effect and pore pressure effect in sand showed that the rate effects in sand might be small, but several references gave contradictory information. The literature on pore water pressure effects is very limited. A research to these phenomena started (Hölscher *et al.* 2008a). This research pointed at answering two fundamental questions:

– The effect of loading rate on the strength of sand and on the bearing resistance of a pile in sand;
– The characteristics of excess pore pressure in sand and in the sand near the pile toe during a rapid load test.

In the laboratory, fast triaxial tests were carried out on dry and saturated sand, in order to determine the constitutive rate effect. The influence of pore water pressures in sand was studied by scale penetration experiments in a calibration chamber.

Numerical studies to the influence of pore water pressure generation during rapid loading have been carried out. This led to a correction formula for the influence of pore water pressure generation.

In order to validate the model for the influence of the pore water under the pile toe, centrifuge tests are carried out. Three cases, which differ in relative density of the sand sample, will be studied at several loading speeds. The piles are loaded with increasing load and in between "static" tests are carried out. The results of these tests are described in (Huy *et al.* 2008; van Tol *et al.* 2008).

Figure 1 Participants at the expert group meeting Delft, 2007.

A more practical study to codes, existing field data and interpretation methods is carried out. A database with field measurements is created. This database contains measured static capacity and derived (from rapid tests) static capacity of piles. The rate effect for sand, silt and clay is estimated. The results of this study are described in (Hölscher, van Tol 2008b).

Finally two types of field tests are carried out:

–   First existing (but never interpreted) data for rapid tests are reinterpreted and compared with static tests. It concerns the results of measurements, which are carried out during the construction of the IFCO building in Waddinxveen.

–   Secondly, a demonstration project is done. Two prefabricated piles are installed at the Waddinxveen site. These piles are tested by a static load test and a rapid load test. The piles are instrumented with a pore water pressure transducer at the pile toe and strain gauges inside.

The results of these tests are included in the database.

3.2   Practical usage of the results

In order to translate the findings of this research a standard for execution of the test and a guideline for interpretation of the test was foreseen. The European unification and increasing number of companies working in international project degrades national standards and guidelines to documents of limited values. A European approach is requested. This approach offers the possibility to integrate the results of all European universities into the documents.

An international expert group has been created. This expert group met in Delft in spring 2004 and spring 2007. The 2004 meeting was an exchange of research results. The 2007 meeting was dedicated to the translation of the results of the research in the documents. The international expert group agreed on the creation of a standard and a guideline for the rapid load test.

Consequently, the working group 4 of TC 341 accepted the standard as part of their task. This means that, after acceptation, the standard will be a formal part of the Eurocode. The national institutes CUR (the Netherlands), BRE (United Kingdom), WCTB (Belgium) and LCPC (France) agreed on cooperation on the creation of a guideline for interpretation of the rapid load test.

4 CONTENTS OF THE STANDARD AND THE GUIDELINE

4.1 Standard for execution

The standard is structured like the similar European standard for static load testing, and the Japanese standard and draft American ASTM standard for rapid load testing were consulted as part of this process.

A dynamic load is considered a rapid load if the loading is first continuously increasing and then continuously decreasing in a sufficiently smooth manner. The load duration must meet the following condition:

$$10 < \frac{T_f}{\dfrac{L}{c_p}} \leq 1000 \tag{1}$$

whereby T_f is the duration of the load, L the pile length and c_p the wave velocity in the pile. This means that the induced wave should be significantly longer than the pile. If the duration of the load is shorter than the prescribed value, the test cannot be considered a rapid load test and should be treated as a dynamic test. If the duration of the load is longer than the prescribed value, the test should be treated as a static test. During the test three variables must be measured continuously:

- the force applied to the pile head
- the displacement of the pile head
- the acceleration of the pile head

The basic requirements for the sample rate and the measurement devices are defined in the Standard. The displacement should be measured relative to a stationary point at sufficient distance from the pile. Before the test, after each loading cycle and after the test an accurate measurement of the pile head level is required as well.

The test should be well prepared and a proper test plan should be drawn up prior to the test in accordance with the minimum requirements included in the Standard. Special attention should be given to site safety aspects.

If there is doubt regarding the integrity of a pile (e.g. due to installation records), an acoustic measurement is prescribed, or the load test must be carried out in multiple steps with increasing pile load. The waiting period after installation equals the requirements for an ordinary static load test.

The obtained measurements must be stored as a function of time on a computer and a back-up medium. Moreover, the rapid load-displacement diagram must be presented as measured and without any correction for mass effects or loading rate.

Finally, the minimum requirements for a test report are defined in the Standard.

A relatively elaborate description of the test execution is required to allow evaluation of the test result by a third party. This will further improve the acceptance of the test results.

The draft standard, which was sent to the TC 341, is added as an annex to this book. It is stressed that the final version must be set up by the technical committee and might divers from this draft.

4.2   Guideline for interpretation

Once the measurements are available, the interpretation must be done. The interpretation of a rapid load test is more complicated than the interpretation of a static load test. Therefore, it is useful to define some methods, which are proven accurate. These will be described in the Guidelines.

The main difference between a rapid load test and a dynamic load test is the duration of the loading. In a dynamic load test, this duration is so short that the stress wave in the pile affects the pile behavior. Due to wave reflections, some parts may be in compression, while others may be in tension. During a rapid load test, this duration is much longer. Consequently the pile moves as one unit, and all pile parts are in a similar state. The wave phenomena in the pile may be neglected during a rapid load test. Nevertheless, during the interpretation of a test, one still must consider the following aspects:

- Influence of the inertia of the pile (expressed as mass multiplied with acceleration);
- Influence of the rate effect;
- Influence of pore water pressures.

The term "rate effect" is used to describe the dependency between the constitutive behavior of a material and the rapidity of loading (Huy, 2008).

In the literature the term rate effect is not uniquely defined, but two descriptions of this effect are meaningful:

- rate effect is the dependency of the constitutive behavior of a material on velocity (rate) of loading;
- rate effect is the dependency of a system on velocity (rate) of loading.

The first description, introduced by Whitman, (1957), considers the rate effect as a constitutive property of a material. It means that stiffness and strength depend on loading rate.

The second description is more general as it includes not only the constitutive property of a material, but e.g. also the damping of a system (due to wave radiation or plastic behavior) and the pore pressure effects. Normally the damping of a system is addressed separately. The damping and inertia effects together might be called the "dynamic effects" in a test.

Excess pore water pressure plays an important role when the soil is undrained or partly drained during loading. In clay, the behavior under a rapid load test can generally be considered as fully undrained. A single-phase description using common material models is applicable. On the other hand, silt and fine-grained sand behave as partly drained material. In these cases the effect of pore water pressures must be handled with care. Finally, in coarse sand and gravel the behavior can be considered as fully drained, and excess pore water pressure does not play a role.

A procedure to calculate the static load displacement curve from the measured data can be written generally as a function of the measured data and some parameters:

$$F_{static}(t) = g[F(t), u(t), v(t), a(t); m, ...]$$ (2)

with $F(t)$ the (measured) force on the pile head
$u(t)$ the (measured) displacement of the pile head
$v(t)$ the velocity of the pile head (calculated from integration of the measured acceleration)
$a(t)$ the (measured) acceleration of the pile head
m the mass of the pile.

Additionally to this function (2), the procedure also contains a description of the method to derive the parameters.

The parametric curve $F_{static}(t)$, $u(t)$ is the derived static load-settlement curve. In the Guideline, the discussion is limited to methods that consider the pile as a single degree of freedom.

The main purpose of the guideline is

– the description of some appropriate functions to describe $F_{static}(t)$ together with the method to derive the parameters for those functions
– the validation of these functions.

One possible interpretation method is the method developed at Sheffield University by Brown, Anderson, Hyde (2004), Brown (2004), which can be described as follows:

$$F_{static}(t) = g_1[F(t), u(t), v(t), a(t); m, \alpha, \beta, v_{static}]$$ (3a)

$$F_{static}(t) = \frac{F(t) - ma(t)}{1 + \alpha \left(\dfrac{v(t)}{v_o}\right)^{\beta} - \alpha \left(\dfrac{v_{static}}{v_o}\right)^{\beta}}$$ (3b)

with m the mass of the pile;
α and β two material dependent model parameters;
v_{static} the speed used to determine α and β;
v_o a normalizing constant.

Table 2 Empirical data of (McVay *et al.* 2003).

Material	Clay	Sand	Silt
Empirical factor	0.53	0.92	0.70

The model parameters α and β are soil dependent and can be measured in a laboratory test. The model holds for positive values of velocity $v(t)$, or the absolute value $abs(v(t))$ must be used instead of $v(t)$. In some articles (e.g. Brown 2008), the parameter α is preceded by a factor proportional to the actual displacement divided by the finally reached displacement.

Another method, known as the Unloading Point Method (UPM), was developed by Middendorp (1992). This method is written as:

$$F_{\text{static}}(t) = g_2[F(t), u(t), v(t), a(t); m, R] \tag{4a}$$

$$F_{\text{static}}(t) = R\,[F(t) - ma(t) - C\,v(t)] \tag{4b}$$

with m the mass of the pile and

R a model parameter that still depends on the soil type.

For the start of the loading the velocity $v(t)$ is small and can be neglected. This leads to a first estimate of the stiffness. In the unloading point (the moment t_u when the displacement is maximum and thus $v(t_u) = 0$), the damping term is zero, and the maximum soil force during the test can be derived. This should be corrected for the rate effect by the factor R, which is material dependant. Table 2 presents the results for the factor R available from literature (McVay *et al.* 2003). Hölscher, van Tol (2008b) presents an extension of this data.

The original unloading point method presents the two main data for the pile: the initial stiffness and the maximum obtained bearing capacity. It does not present the full-derived load-displacement curve, since the parameter C_4 is unknown. Several methods for estimation are presented in literature.

A third method is presented by (Matsumoto *et al.* 1994), which is also discussed in (Matsumoto 2008). The numerical integration method contains of a number of sets with two steps. In the odd steps, the dashpot constant is updated, and in the even steps, the stiffness is updated. The derived static force is calculated in each step. For the full equations, we refer to the equations (7) to (15) in the publication of (Matsumoto 2008). Since the method of Matsumoto is based on similar equations as UPM (but with a non-constant dashpot constant C), we expect that a similar model factor R must be applied in this method.

4.3   Field data

The guideline will be accompanied with a collection of field data. Field data plays an essential role in the validation and reliability analysis of the rapid load test. Many authors pay attention to field data.

(Huybrechts *et al.* 2008) summarize the results of lessons from the two field tests at Limelette and Sint-Katelijne-Waver in Belgium. Several types of piles were tested both static and rapid. Based on these field tests they conclude that the safety factors for a rapid load test cannot be equal to the safety factors for a static load test. For clay they hesitate for the applicability of a rapid load test as a design test.

(Brown 2008) stresses the non-linear behavior of clay during the rapid load test. The analysis of a rapid load test on a pile in clay is much more complex than in sand. He also doubts on the possibility to derive a bearing capacity of a pile in clay by a rapid load test.

(Hyde and Brown 2008) studies the load transfer in a pile in clay during a rapid load test. While the stiffness behavior during a rapid load test is almost identical to that during a static load test, the bearing capacity needs a much higher correction. They confirm the factor 2, mentioned in Table 2.

5 CONCLUSION

The Delft Cluster research started with some hesitation on the applicability of a rapid load test for pile foundation testing. The actual development of European unification and increasing number of companies working in international project forced the project to work in an international working field. Due to the international collaboration, it turned out to be possible to do within a limited period:

- fundamental research on dynamic soil behavior for both sand and clay
- collecting a good number of measurement data
- writing a standard and a guideline, which can be applied in practical cases.

In collaboration with all partners, the world of foundation design made a big step forwards.

ACKNOWLEDGEMENT

All parties who were involved in this project are kindly acknowledged.

REFERENCES

Bermingham, P.D. and Janes, M.C. (1989). An innovative approach to load testing of high capacity piles, Proc. Int. Conf. on piling and deep foundations, pp. 409–413.

Brown, M.J., Anderson, W.F. and Hyde, A.F. (2004). Statnamic test-ing of model piles in a clay calibration chamber; In: Int. Jnl. Phys. Modelling Geotechnics., Vol. 4, No. 1 pp. 11–24 (IS-SN 1346-213X).

Brown, M.J. (2004). The rapid load testing of piles in fine grained soils, thesis University of Sheffield, Dep. of Civil Eng. and Structural Eng., March, 2004.

Brown, M.J. (2008). Recommendations for Statnamic use and interpretation of piles installed in clay, Rapid Load Testing on Piles, Hölscher, P., van Tol, A.F. (eds), Francis Taylor, September 2008.

Gonin, H., Coelus, G. and Leonard, M. (1984). Theory and performance of a new dynamic method of pile testing, proc. 2nd int. Conf. Application of stress-wave theory to piles, stockholm: 403–410.

Hölscher, P. www.rapidloadtesting.eu, 2007.

Hölscher, P., van Tol, A.F. and Huy, N.Q. (2008a). Influence of rate effect and pore water pressure during Rapid Load Test of piles in sand, Rapid Load Testing on Piles, Hölscher, P., van Tol, A.F. (eds), Francis Taylor, September 2008.

Hölscher, P. and van Tol, A.F. (2008b). Database of field measurements of SLT and RLT for calibration, Rapid Load Testing on Piles, Hölscher, P., van Tol, A.F. (eds), Francis Taylor, September 2008.

Huy, N.Q., van Tol, A.F. and Hölscher, P. (2008a). Rapid Model pile load tests in the geotechnical centrifuge, part 2, Rapid Load Testing on Piles, Hölscher, P., van Tol, A.F. (eds), Francis Taylor, September 2008.

Huybrechts, N., Maertens, J. and Holeyman, J.A. (2008b). Belgian national project on screw piles – Overview of the comparative load testing program, Rapid Load Testing on Piles, Hölscher, P., van Tol, A.F. (eds), Francis Taylor, September 2008.

Hyde, A.F.L. and Brown, M.J. (2008). Load Transfer in Rapid Load Pile Tests in Clays, Rapid Load Testing on Piles, Hölscher, P., van Tol, A.F. (eds), Francis Taylor, September 2008.

Janes M.C., Bermingham, P.D. and Horvath R.C. (1991). An innovative dynamic test method for piles, Proc. 2nd Int. Conf. Recent advances in geotechnical earthquake engineering and soil dynamics, St. Louis, Missouri, Paper No. 2.20, 1991.

Matsumoto, T. (2008). Practice of rapid load testing in Japan, Rapid Load Testing on Piles, Hölscher, P., van Tol, A.F. (eds), Francis Taylor, September 2008.

Middendorp, P., Bermingham, P. and Kuiper, B. (1992). Statnamic load testing of foundation piles. 4th International Conference on Stress Waves, The Hague, Balkema, 1992.

Schellingerhout, A.J.G. and Revoort, E. (1996). Pseudo static pile load tester, Proc. 5th Int. Conf. Appl. Stress-Wave Theory to Piles, Orlando, Sept. 1996, pp. 1031–1037.

Stokes, M. and Mullins, G. (2008). Concrete Stress Determination in Rapid Load Tests, Rapid Load Testing on Piles, Hölscher, P., van Tol, A.F. (eds), Francis Taylor, September 2008.

Van Tol, A.F., Huy, N.Q. and Hölscher, P. (2008). Rapid Model pile load tests in the geotechnical centrifuge, part 1, Rapid Load Testing on Piles, Hölscher, P., van Tol, A.F. (eds), Francis Taylor, September 2008.

Whitman, R.V. (1957). The behaviour of soils under transient loadings, Proc. Fourth Int. Conf. on Soil Mechanics and Foundation Engineering, pp. 207–210.

Seminar papers: Advances in international research

Belgian national project on screw piles – Overview of the comparative load testing program

N. Huybrechts
Belgian Building Research Institute (BBRI), Brussels, Belgium

J. Maertens
Jan Maertens bvba & Catholic University of Leuven (KUL), Belgium

A. Holeyman
Catholic University of Louvain, Louvain-La-Neuve (UCL), Belgium

SUMMARY

The Belgian Building Research Institute (BBRI) organized, with the financial support of the Belgian Federal Ministry of Economical Affairs, a research project concerning ground displacement screw piles (BBRI 1998–2000 & 2000–2002). This contribution gives a general overview of the research program, and the different load test types that have been performed.

The authors express their points of view regarding the application of kinetic load test and the establishment of a guideline and/or a test standard.

1 INTRODUCTION

The *"ground displacement screw pile"* is a Belgian technology, of which the market share has increased enormously over the last years and which is still increasing. Also on an international level the interest is growing. This success can partially be explained by the ground displacement characteristics of these piles (no soil removal) and their high installation speed. On the other hand the vibration-free and the low-noise installation method play a very important role, especially in densely populated and urban areas.

In order to calibrate the semi-empirical calculation methods, which are mostly based on CPT tests in Belgium, to investigate more in detail the behaviour of this pile type, and to apply and analyse alternative (and cheaper) test methods to deduce the static pile behaviour, i.e. dynamic and kinetic load testing, the BBRI carried out a major research project addressing cast in-situ ground displacement screw piles during the period 1998–2002 (BBRI 1998–2000 & 2000–2002).

The project took place with the financial support of the Belgian Federal Ministry of Economical Affairs and was carried out in collaboration with five Belgian piling companies: De Waal Palen, Franki Geotechnics B, Fundex, Olivier and Socofonda. A National Advisory Committee under supervision of prof. A. Holeyman (UCL) and prof. J. Maertens (KUL) guided the research program.

2 GENERAL REVIEW OF THE RESEARCH

In the first stage of the project (BBRI, 1998–2000) 5 types of screw piles and driven precast piles were installed on a site in Sint-Katelijne-Waver (B) where the subsoil consists of O.C. tertiary Boom clay. Pile loading tests were executed on 30 test piles: 12 static load tests, 2 series of twelve dynamic load tests and 6 Statnamic tests. The results of this test campaign were extensively reported during the first symposium *"Screw piles – Installation and Design in Stiff Clay"*, which was held on 15 March 2001 in Brussels. The proceedings of this symposium have been published in English by Swets & Zeitlinger (Balkema), ISBN 90 5809 192 9 (editor A. Holeyman).

In the second stage (2000–2002), a test campaign of similar extent was organised on a site in Limelette (B), where the subsoil consists of quaternary silty layers (loam) and tertiary Ledian-Bruxellian sand. The results of the second test campaign were reported during a second symposium *"Screw Piles in Sand – Design and Recent Developments"* that took place on 7 May 2003. The proceedings of this 2nd symposium have as well been published in English by Swets & Zeitlinger (Balkema), ISBN 90 5809 578 9 (editors J. Maertens & N. Huybrechts).

Both volumes contain all details about the test campaign (geologic background, soil investigation program, test results, outcome of an international prediction event, ...)

A typical CPT of both test sites is given in Figure 1.

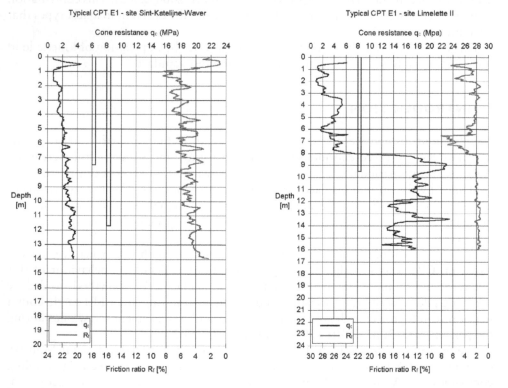

Figure 1 Typical CPT E1 on the site Sint-Katelijne-Waver (left) and Limelette (right).

3   DESCRIPTION AND RESULTS OF THE RESEARCH

On both test sites, different types of load tests were performed: static load tests (maintained load test procedure), dynamic load tests and kinetic (Statnamic) load tests.

In order to compare the results of the static load tests with the static behaviour deduced out of analysis of the dynamic and kinetic load tests, an international prediction event was organised. The details of this comparative analysis was published in contributions by Holeyman *et al.* (2001 & 2003) in the previously mentioned volumes.

Figure 2 and 3 illustrate an example of the results of that prediction event for the clay and sand site respectively.

With regard to the results of the kinetic load tests the following general conclusions were reported:

For the *clay* site, the Statnamic (STN) predictor used the Unloading Point Method (UPM) to predict the static load-settlement behaviour. It was mentioned by the predictor that, due to strain rate sensitivity of clayey soils, a 30% reduction coefficient had to be applied on the usual UPM method. A hyperbolic approximation of that reduced function was then calculated. This is the reason why those predictions are reported "0.7 STN" (see figure 2). Even with this reduction of 30% the "0.7 STN" prediction overestimated the ultimate pile capacity (defined as the load corresponding with a pile displacement of $10\%D_b$) by 25% in average, which means that the results obtained by the UPM overestimate pile capacity by ±50%.

This clayey soil seem extremely sensitive to pile rate effects. The generated pile velocity during the kinetic load tests seems the governing factor that determines the mobilised pile resistance. Applying a simple reduction factor in order to fit ultimate load, is not an acceptable approach as it would generate deviations of the load settlement curve in the initial part of the curve (working load range).

It should also be remarked that in these clay layer even different procedures applied for static load testing (duration of load steps) might significantly influence the static pile capacity deduced from these tests.

For the *"dry" sand* site (the sand was not actually dry, since it was in the vadose zone, but the sand is referred herein as dry for simplicity, while indicating it was not located below the water table) the mobilised static resistance deduced by the Statnamic (STN) predictor was in average 11% higher than the resistance mobilised during static load tests, and this for a pile head settlement of 10 mm ($2.5\%D_b$). This overestimation apparently corresponded well with the expectations of the STN predictor (influence strain rate effects).

With regard to the pile load tests campaigns at Sint-Katelijne-Waver and Lime-lette, it is especially the results of the static pile load tests that have been extensively exploited within the framework of the establishment of the Belgian National Annex of the Eurocode 7. Up to know the results of the kinetic load tests have not been analysed further in detail by BRRI.

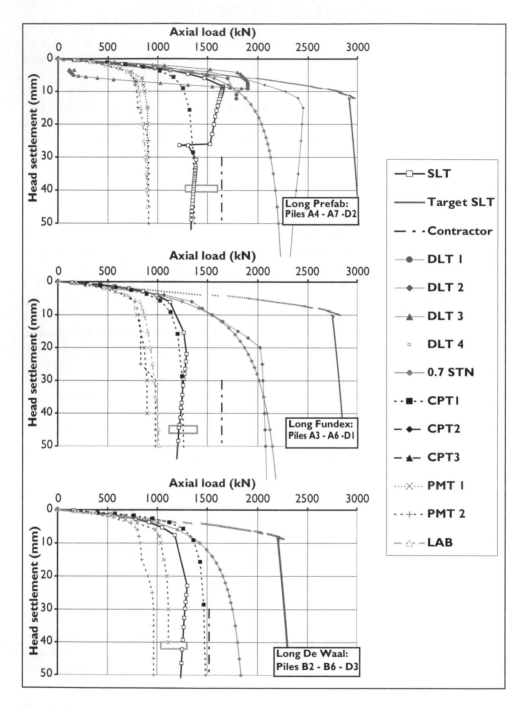

Figure 2 Examples of the result of the prediction event at the clay site (Sint-Katelijne-Waver).

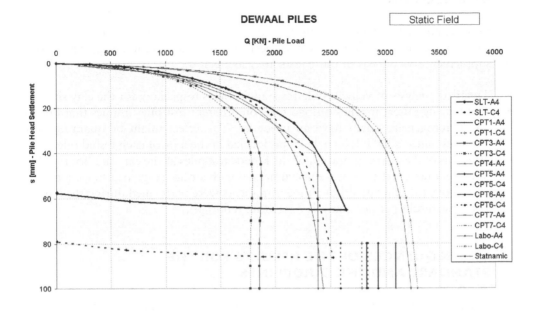

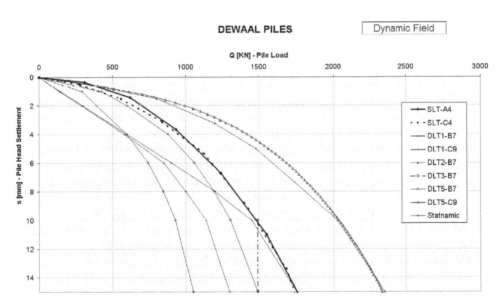

Figure 3 Example of the result of the prediction event at the sand site (Limelette).

4   CONCLUSION

Up to now, no further detailed analysis of the results of the kinetic load tests in Sint-Katelijne-Waver and Limelette have been performed. Some (non-limiting) suggestions for supplementary analysis of these tests are listed below:

- Detailed analysis of velocity-pile resistance effects (especially for the clay site)
- Comparing different types of load tests on different test piles implies that (local) site heterogeneity might have an influence. This effect might be quantified for both test sites as CPT-E have been performed in the axis of each tested pile.
- Analysis of the instrumentation of the kinetic test piles at the clay site (for the clay site the kinetic test piles were instrumented with strain gauge transducers placed just above the pile tip; although measurements were performed during the kinetic tests, no results neither analysis have been reported).

5   CONSEQUENCES OF THE RESULTS FOR THE STANDARD AND THE GUIDELINES

With regard to the application of kinetic load test and the establishment of a guideline and/or a test standard, the authors wish to put forward the following (Belgian) viewpoint and requirements:

Kinetic load tests in general

- Soil mechanics research at high strains and granular matter computational physics show that rate dependency is neither monotonous, nor reversible.
- Avoid single blow RLT: site-specific experimental derivation of displacement, velocity, and acceleration dependence of results needs to be ascertained: demonstrate stability of prediction through multiple blows tests.
- An extensive and demanding test procedure (minimum number of blows, minimum pile head displacement, ...) to perform kinetic load testing on piles is necessary. The test procedures need to be standardized taking into account the aim of the test (either control, design, or research).
- The influence of the load test procedure (a.o. number of blows) needs to be studied and more widely ascertained. Especially in clayey subsoil a possible influence of the number of cycles (pore water pressures) is to be considered.
- Comparative load test procedures (static versus kinetic) have to be encouraged within a probation period. When or under which conditions would RLT become a single reliance for pile acceptance. How should we deal with different type of load tests on the same test pile (how can interference between different load tests be mitigated?), and/or how should we deal with different tests types performed on separate test piles (local heterogeneity might interfere with direct comparison). Also the existence of different static load test procedures, which might influence the 'reference' static load settlement behaviour, should be taken into account.
- For comparative analysis, the static pile behaviour deduced from kinetic load testing should only be compared with the results of static load tests (see also

remarks previous point) and not (only) with semi-empirical design methods based on CPT e.g.

Kinetic load test as control test

- Kinetic load testing seems to have a potential to be applied as an alternative for static pile load tests (up to 1.5 design load).
- More extended investigation is however needed to perform comparative analysis in the working load range and this for different soil types (especially in cohesive soils and in saturated soils).
- In certain conditions (e.g. specific subsoil conditions, no comparable experience available, considerable variation of the results, ...) a higher load mobilization should be enforced (say 2 or more times the design load), in order to cover some additional uncertainty.

Kinetic load test as design test

- One of the methods that is allowed by Eurocode 7 for pile design, is the design bases on preliminary load tests. The design load is then deduced from the test results (in principle the ultimate static pile load) by applying ξ factors (depending on the number of tests, stiffness of the structure) and γ factors (partial safety factors).
- It is the authors' opinion that these factors (especially the safety factors) can not be the same as the factors to be applied for static load tests. These factors need to be reviewed in detail, based on extended comparative load test data in different soil types. Possibly an extra safety (model) factor should be integrated in this procedure.
- Based on the actual available data resulting from tests in Sint-Katelijne-Waver, the application of the kinetic load as design test in clayey subsoil seems for the moment to be highly questionable. Also for saturated sand no comparative analysis is available in Belgium until now.

REFERENCES

BBRI (1998–2000 and 2000–2002). Soil displacement screw piles – calibration of calculation methods and automatisation of the static load test procedure: stage 1 – friction piles & stage 2 – end-bearing piles. Research programs subsidised by the Belgian Federal Ministry of Economical Affairs, convention numbers CC-CIF-562 and CC-CI-756.

Holeyman, A., Couvreur, J.M. and Charue, N. (2001). Results of dynamic and kinetic pile load tests and outcome of an international prediction event, Proceedings of the symposium Screw piles – installation and design in stiff clay, March 2001, Brussels.

Holeyman, A. and Charue, N. (2003). International pile capacity prediction event at Limelette, Proceedings of the 2nd symposium on screw piles, May 2003, Brussels.

Middendorp, P. (1999). Statnamic test results Sint-Katelijne-Waver. Profound Report 99-PROF-R0014 (Draft version).

Middenddorp, P. and Keindorf, C. (2001). Statnamic load testing – 2nd phase site of the project in Limelette. Profound Report 2001-PROF-R0012.

Swets and Zeitlinger (2001). Holeyman, A. (ed.). Screw Piles: installation and design in stiff clay, Lisse (NL).

Swets and Zeitlinger (2001). Maertens, J. and Huybrechts, N. (eds.). Belgian screw pile technology – Design and recent developments, Lisse (NL).

Chapter 2

Recommendations for Statnamic use and interpretation of piles installed in clay

M. Brown
University of Dundee, UK

SUMMARY

The following presents guidance for the effective use and analysis of Statnamic testing, based upon five years of research undertaken at the University of Sheffield and Dundee. The major focus of this research has been the development of non-linear strain dependant analysis techniques for piles installed in clay. This has lead to the development of simple analysis that requires soil specific damping parameters as input. This technique has been found to perform well for skin friction piles in both glacial till and London Clay.

Undertaking both laboratory and field studies has given insights into potential improvements in both testing techniques and equipment. The most important of these is the need for users to apply loads significant enough to achieve maximum pile settlements equivalent to 10% of pile diameter. Thus, tests should not be terminated on maximum load criteria but more appropriately on maximum and permanent settlement achieved. This is to improve both analysis and to avoid selection of Statnamic devices incapable of applying adequate loads. As a rough guide, in clay soils the Statnamic device should be capable of applying loads twice the predicted ultimate static capacity. This should preferably be applied as a single loading event rather than cycling load pulses.

Accuracy of the Statnamic test should be verified through the use of high-precision optical levelling before and after all loading events. Both analysis and settlement measurement would benefit from the incorporation of an accelerometer at the head of the pile. This installation should be standardised with minimum logging rates stated. In addition, non-contact settlement measurement should be undertaken with respect to the underlying soil conditions with the pile-measurement device separation being selected as appropriate.

1 INTRODUCTION

For Statnamic pile testing to become more widespread, consultants and contractors alike need to be confident that the technique can be used to predict static pile behaviour in any ground conditions or soil types. Analysis and deployment of the technique in coarse grained soil types has proven relatively straight forward. As with early experience in dynamic testing, it was quickly realised that the main barriers to widespread Statnamic deployment lay testing in fine grained soils such as clay and silt. In 1999 the University of Sheffield embarked on a series of research studies with the aim of investigating the issues involved with Statnamic testing in clay. The main aims of the projects were to improve general understanding of rapid load pile testing in these soil types and develop appropriate analysis techniques. This was achieved through laboratory and full scale testing. Since 2003 work has continued in Sheffield and Dundee in the form of further laboratory studies and the analysis of case study information with the predominant focus on improving analysis techniques.

This paper summarises the previous and current research work into test analysis and reports the latest techniques being used. In addition to the above the paper looks at techniques for more effective field testing and deployment based upon past experience and recent analysis developments. The findings of the research to date are summarised and used to make suggestions for incorporation in standard codes of practice.

2   GENERAL REVIEW OF THE RESEARCH

2.1   Recently completed research

The author worked on two EPSRC funding projects between 1999 & 2003. The first project involved three years of testing a fully instrumented model pile at various strain rates and under simulated Statnamic load pulse (Brown *et al.* 2004; Brown 2004). The outcomes of which included proving the validity of applying a previously developed non-linear velocity dependant rate effect analysis technique at Statnamic type strain rates in clay (Brown 2004). In order to verify the findings of the three year laboratory study, a full scale test site was developed (in glacial till, Grimsby, UK) which included a precast driven pile and a heavily instrumented auger bored pile. In addition, the test facility included soil instrumentation in the form of a series of radial accelerometers arranged around the auger bored pile (Brown *et al.* 2006; Brown & Hyde 2008). Statnamic testing was undertaken on both piles with cycles of load of 1000, 1500, 2000, 2500 and 3000 kN applied to the auger bored pile followed by CRP and MLT static testing. The results of this study were used to provide guidance for analysis, specification, on-site working practices and equipment developments (Brown & Hyde 2006; Brown *et al.* 2006). It should be noted that these two projects focused significantly on understanding skin frictional pile resistance rather than the end bearing resistance.

2.2   Current research

Current research at the University of Dundee is based upon testing previously developed concepts and observations on case study examples when these become available (Powell & Brown 2006). The focus is on improved analysis in fine grained soils using the non-linear damping parameter based approach but with the inclusion of simple strain dependant damping, as proposed by Balderas Meca (Balderas-Meca 2004). Case studies are being used to investigate the validity of cyclic Statnamic testing, the effect of pile mobilisation on the performance of analysis and the use of soil characteristics to predict strain rate dependant behaviour (Powell & Brown 2006; Brown & Hyde 2006).

2.3   Future research

The geotechnics research group at the University of Dundee has embarked on a three year funded research project commencing September 2007 designed to look specifically at rate effects in fine grained soils. The project will examine the behaviour of fine grained soils over a wide range of strain rates whilst varying soil composition. This approach will identify both behaviour at key strain rates and allow understanding of what the main controlling factors are at soil micro structural level. Laboratory testing will include high speed monotonic triaxial testing with on-sample strain and pore pressure measurement. In parallel, the soils will be characterized using simple standardized laboratory testing techniques. The high strain rate testing and standard testing will then be compared to develop a predictive framework which will allow the determination of soil rate potential from standardised laboratory tests, without the need for specialised testing or empirical studies. The predictive framework will be of significant benefit to rapid load pile testing. The need for this research has been identified through the author's previous studies (Powell & Brown 2006) and the attempts of others to develop soil specific rate parameters linked to measurable properties such as moisture content and plasticity index (Gudehus 1981; Powell & Quaterman 1988).

3   DESCRIPTION AND RESULTS OF THE RESEARCH

The initial part of this section focuses on the analysis of Statnamic tests based upon both laboratory and field testing experience. The latter part focuses on observations of field testing techniques.

3.1   Analysis procedures for friction piles in clay

Results from previous research have shown that it is possible to deduce the equivalent ultimate static pile resistance from Statnamic tests using a non-linear velocity dependant model (Eq. 1)(Brown *et al.* 2006). This form of analysis is based upon that proposed by Randolph & Deeks (Randolph & Deeks 1992) for analysis of dynamic tests where the majority of the pile capacity is developed through skin friction.

$$F_{static(ultimate)} = \frac{F_{STN} - (M\ddot{x})_{pile}}{1 + \alpha(\Delta v / v_0)^\beta - \alpha(\Delta v_{min} / v_0)^\beta} \tag{1}$$

Where F_{static} is the static pile resistance, F_{STN} is the measured Statnamic load, $M\ddot{x}$ is the pile inertia, Δv is the pile's velocity relative to the soil and v_{min} is the velocity of the static CRP pile test used to define the soil specific rate parameters α and β. The value of α was taken as 1.26 (Table 1) as previously determined from laboratory testing. The parameter β is normally set to a value of 0.2 for clay soils (Randolph & Deeks 1992). Where analysis has been undertaken for London Clay (Powell & Brown 2006) a value of α equal to 1.77 was used as per table 1.

The performance of this approach (Eq. 1) applied to the results from the Grimsby field study are shown in figure 1. The predicted static capacity of 1746 kN was only

Table 1 Summary of rate parameters from previous studies (Powell & Brown 2006).

Originator	Soil type	Index properties (LL, PL, PI-%)*	α	β	Test conditions
Randolph & Deeks	Sand	–	0.1	0.2	Summary of previous work
	Clay	–	1.0	0.2	
Balderas-Meca	Grimsby glacial till	20–36, 12–18, 7–20	0.9	0.2	Full scale Statnamic tests
Brown	Model clay	37, 17, 20	1.26	0.34	Model Statnamic tests
Litkouhi & Poskitt	London clay	70, 27, 43	1.77	0.18	Model pile skin friction tests
	Forties clay	38, 20, 18	0.99	0.23	
	Magnus clay	31, 17, 14	0.86	0.46	

* Ll-Liquid Limit, PL-Plastic Limit, PI-Plasticity Index.

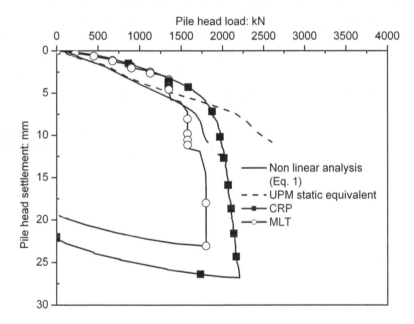

Figure 1 Comparison of derived ultimate static equivalent pile behaviour with measured static CRP & ML tests (Brown *et al.* 2006).

10% less than the measured CRP load of 1946 kN. In comparison, the unloading point method (UPM)(Kusakabe & Matsumoto 1995) over predicts the pile capacity by 31% and fails to define a clear yield point (Brown *et al.* 2006). Application of the approach in equation (1) to the pre-yield phase of loading resulted in an under prediction of stiffness which was also true for the UPM approach. This is not surprising as the empirical rate parameters used were based upon ultimate model pile studies with no regard for the pre-yield phase (Brown *et al.* 2006). In addition, the applicability of this approach to the pre-yield phase is questionable, based upon field observations of

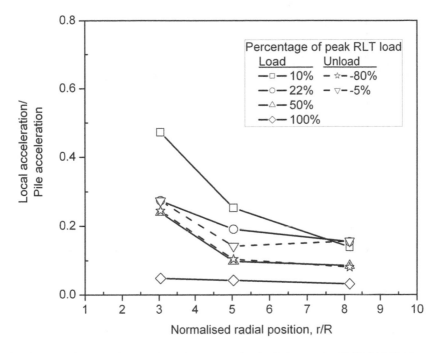

Figure 2 Influence of pile loads on soil acceleration at 4 m BGL during the 3000 kN RLT (Brown *et al.* 2006).

soil accelerations which show decoupling of pile accelerations from those of the surrounding soil on approaching peak load (Fig. 2).

In order to expand this form of analysis to both working and ultimate pile behaviour, Balderas-Meca (Balderas-Meca 2004) suggested making the value of the rate parameter α dependant on the level of pile-soil strain with linear variation of α up to approximately 1.0 to 1.2% of the pile settlement relative to the pile diameter. Above this level, consistent with ultimate pile behaviour, the value of α becomes constant (Fig. 3). To incorporate this variation in α in a simple and user-friendly form, equation (1) was modified by Brown (Brown *et al.* 2006) to:

$$F_{static} = \frac{F_{STN} - (M\ddot{x})_{pile}}{1 + (\frac{F_{STNs}}{F_{STNpeak}})\alpha(\frac{\Delta v}{v_0})^\beta - (\frac{F_{STN}}{F_{STNpeak}})\alpha(\frac{\Delta v}{v_0})^\beta} \tag{2}$$

Where it is assumed that at peak Statnamic load ($F_{STNpeak}$) the pile has been significantly mobilised and that the value of α has become constant. This is a prerequisite for accurate analysis using both equation (1 & 2). Results of analysis using equation (2) are shown in figure 4 compared with UPM modified by a factor of 0.65 as suggested by Paikowsky (Paikowsky 2004) for testing in clays. When corrected by 0.65 the UPM prediction of ultimate behaviour is significantly improved but the prediction of pile stiffness is made worse. This suggests that the fixed damping constant approach

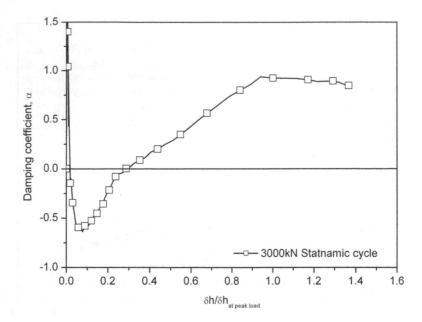

Figure 3 Settlement dependency of rate parameter α (Powell & Brown 2006).

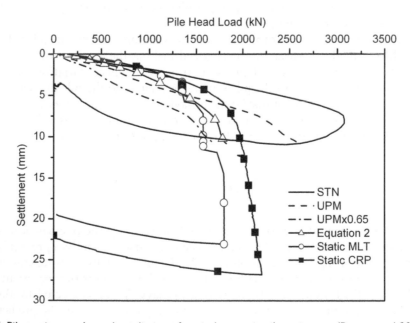

Figure 4 Pile testing results and prediction of equivalent static pile resistance (Brown *et al.* 2006).

used by UPM is inappropriate in the pre-yield zone when analysing tests in clay. Analysis using the non-linear approach shows improved prediction of both ultimate static resistance and pile stiffness although predicted pile stiffness is still conservative (Brown *et al.* 2006).

Both the methodology in equation (2) and the suggested improvements to UPM require user intervention through the selection of either soil specific rate parameters or correction factors. The use of such factors detracts from the original ethos of UPM which was developed to be based upon measured data without the need for user input (Brown 2008). Currently parameters specific to Statnamic velocities are empirically derived based upon very limited data set in clays. A summary from selected previous studies of values for non-linear analysis is given in table 1 (Powell & Brown 2006). The use of the term "clay" as a catch-all for selection of a rate parameter does not reflect the subtleties of the material where it has been shown previously that rate effects may vary significantly in the same material at different void ratios (Brown & Hyde 2006; Gudehus 1981; Powell & Quaterman 1988).

It is interesting to note that from Table 1 that the magnitude of α in clay soils appears to reduce with reduction in plasticity index (PI) such that (Powell & Brown 2006):

$$\alpha \approx 0.03PI + 0.5 \tag{3}$$

However it should be noted that this relationship is very tentative at present and requires further validation. In addition, use of the non linear approach presented in equation (2) should be restricted to piles developing the majority of their capacity from skin friction. Use of this approach where end bearing is a significant factor will lead to under predictions of static capacity (Powell & Brown 2006). Where soil specific damping parameters do not exist for a clay these should be determined through correlation with static pile testing in the same soil or through high strain rate laboratory testing.

Adopting the variation of α with increasing pile settlement shown in figure 3 highlights the need for adequate settlement of the test pile to allow accurate analysis. If the pile has not reached a significant displacement corresponding to ultimate behaviour, (and a maximum α value) it is difficult to select an appropriate final damping constant for inclusion in equation (2)(Powell & Brown 2006). This comment is also true for UPM analysis where it is assumed at the unloading point that the pile has achieved its equivalent ultimate static pile resistance. From a testing point of view this raises questions as to whether or not it is appropriate to apply a single loading event designed to mobilise the full shaft resistance or to carry out a series of cycles of ever increasing loading (Powell & Brown 2006). This question is further complicated by the results shown in figure 5 where it is apparent that the magnitude and rate of change of α is influenced by the magnitude of Statnamic load applied and that the α settlement relation shows characteristics that are apparently an artifact of the stress controlled nature of the test.

From an analysis view point, based upon the currently used techniques i.e. UPM and non-linear, it would seem inappropriate to use cyclic loading approaches. Additionally, where cyclic approaches are used to derive rate parameters and verify analysis there is the potential for strain hardening (or softening) which will effect prediction and back analysis, especially in the pre-yield zone of pile behaviour (Brown *et al.* 2006).

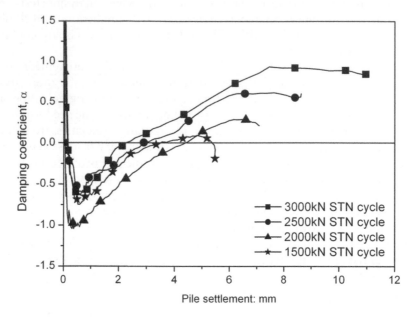

Figure 5 Settlement dependency of rate parameter α at different load levels during cycles of Statnamic loading.

3.2   Some observations on Statnamic pile testing

As Statnamic is a short duration high stress event it is not surprising that the test effects may be felt some distance from the test area. As the use of a remote laser reference source (or optical measuring system) for settlement measurement is typical, careful consideration should be given to pile-laser separation. The results of interference of the Statnamic device with the laser reference source can be clearly seen in figure 6 (Brown & Hyde 2006). Between points B and C the settlement suddenly increases at the end of load application. As no such displacement was apparent from the accelerometers embedded in the pile, this shift was attributed to movement of the laser reference source. Analysis of S-wave travel times to the laser also showed that this was feasible and the author can verify that significant ground vibrations were felt during testing. To avoid surface wave induced vibrations interfering with settlement measurements during the load pulse, it is recommended that the laser is placed at adequate distance from the pile chosen based upon the encountered surface soil type (Table 2).

The earlier developments of the Statnamic device did not incorporate the facility to measure acceleration directly. Both velocity and acceleration being obtained from differentiation of the settlement-time history. The process of differentiation tends to amplify signal noise, and it is therefore necessary to smooth the data. This process may lead to a reduction in peak accelerations especially where logging of the Statnamic measurements is undertaken at low rates (Brown & Hyde 2006). The improvements associated with using an accelerometer to calculate velocity are clearly shown in figure 7.

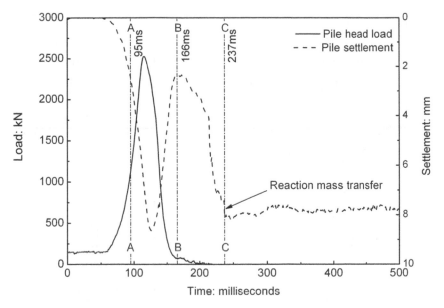

Figure 6 Measured load and pile settlement during a 2500 kN Statnamic pulse (Brown & Hyde 2006).

Table 2 Recommendations for pile-laser source minimum separation for varying soil types (Brown & Hyde 2006).

Soil type	Typical S-wave velocity: m/s	Minimum laser-pile separation: m
Loose sand	90–155	16
Dense sand	230	23
Soft to firm clay	155	16
Stiff clay	210	21
Loose gravel	185	19
Dense gravel	230	23

Although the selection of load magnitudes has been mentioned above, it is worth reiterating that Statnamic tests should be used to create significant settlement in piles. The author is often asked to analyse tests where settlements are less than or equal to those anticipated at working loads and often encounters specifiers that were surprised when they could not prove equivalent ultimate static capacity with the selected Statnamic device (Fig. 8). This is predominantly due to the specification of maximum test loads based upon recommendations for static testing i.e. 100% SWL + 50% SWL (safe working load) but without the realization that where rate effects are significant, this magnitude of load may only be equivalent to 50 to 75% SWL. Thus to aid analysis, it has previously been suggested that testing should be terminated on settlement

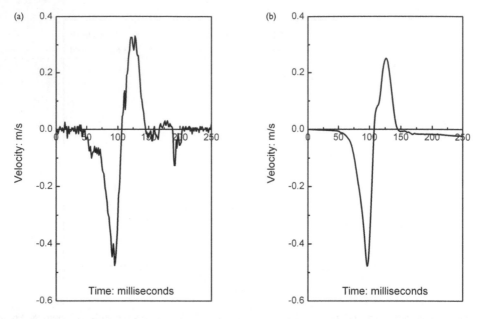

Figure 7 Comparison of pile velocity (a) calculated from measured settlement-time history and (b) calculated from measured pile acceleration at 4 m BGL during a 3000 kN Statnamic test (Brown & Hyde 2006).

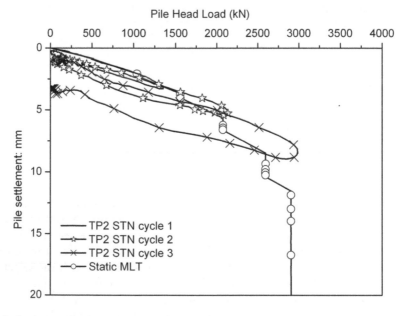

Figure 8 Statnamic results for a 27 m long 450 mm diameter auger bored pile installed in London Clay where the maximum applied Statnamic load is approximately equal to the ultimate static capacity (Brown *et al.* 2006).

rather than load criteria with minimum settlements of 10% of the pile's diameter being recommended. For planning Statnamic tests a device should be selected that can apply at least twice the ultimate static capacity with minimum loads applied during Statnamic equivalent to 1.7 to 1.75 times the ultimate static capacity (Brown & Hyde 2006; Brown et al. 2006; Powell & Brown 2006).

The above material is based predominantly on the testing of 12 m long, 600 mm diameter auger bored piles in a glacial till (Brown 2004; Brown et al. 2006; Brown & Hyde 2006). These findings have also been verified for testing with a 1000 kN Statnamic device on 300 mm diameter auger bored piles installed in London Clay (Brown et al. 2006; Powell & Brown 2006).

4 CONSEQUENCES OF THE RESULTS FOR THE STANDARD AND THE GUIDELINES

Currently Statnamic is mentioned in several national guidance documents. For instance in the UK the forthcoming revision to the The Institution of Civil Engineers specification for piling and embedded retaining walls (SPERW) will mention rapid load testing in its non-destructive testing section. The Federation of Piling Specialists (http://www.fps.org.uk/) has recently published simple guidance on use of the technique as part of a general pile testing guide. In the US a draft testing specification has been produced for the ASTM standard (Janes et al. 2000) along with guidance produced for the federal and state highways agencies (Paikowsky 2004; McVay et al. 2003). The Japanese Geotechnical Society have also produced a testing specification, an English language draft of this was published in 2000 (The Japanese Geotechnical Society 2000). In addition to this guidance, it is the author's opinion that any future guidance should address the following issues based upon previous and current research findings. These points are made for the testing of skin friction piles installed clay soils. Where other soil types and pile configurations exist these recommendation will require further validation.

4.1 Analysis

Analysis of the obtained Statnamic data is relatively straight forward in granular soils and piles with rock sockets. In fine grained soils such as clay and silt more specialized non-linear techniques are needed which require manual input of soil dependant rate parameters. In order for such analysis to be successful it is important to mobilise significant pile settlement during Statnamic loading which will involve the application of loads at least 1.75 times the static ultimate resistance in clay soils.

The use of a non-linear analysis technique with damping parameters coupled to pile displacement has been shown to perform well using rate parameters derived from previous studies. This method appears to give more reliable prediction of equivalent ultimate static pile resistance in London Clay than the existing unloading point method. It has also been shown to provide more accurate prediction of soil-pile stiffness in glacial clays.

It should be noted, however, that the proposed non-linear analysis and rate parameters were derived from piles developing the majority of their capacity from skin friction and further verification will be required where pile end bearing capacity is significant.

4.2   Statnamic loading strategies

In terms of a Statnamic loading strategy and device selection, the target loads should be selected to produce pile settlement similar to that considered a minimum to define ultimate load during static pile testing. (e.g. 10% of the pile's diameter). Where the Statnamic device is used to produce settlement that is normally associated with working rather than ultimate loads, the analysis must be calibrated directly against test pile results (i.e. static CRP) installed in similar conditions. Care should be taken when specifying the Statnamic device that it can operate over the loading range required to produce adequate settlement.

Based upon the current development of analysis techniques, it is suggested that cyclic loading with the Statnamic device is avoided where the aim of testing is to derive the equivalent static ultimate pile resistance and the load–settlement curve to that point. Ideally a single load application approach should be used with the test termination criteria being maximum settlement achieved rather than a factor of safe working load. It is appreciated that this approach is not straight forward especially where there are uncertainties about the rate dependant nature of the soil. A simple approach to estimating target load in clays would be to apply twice the anticipated ultimate static load.

4.3   Site and equipment practicalities

For all pile load-settlement determinations it should be mandatory for careful optical levelling of the pile head elevation prior to and after each cycle of loading. This is made more important through the use of a remote displacement measuring system that may be influenced by the pile testing methodology. In addition these remote systems are typically not linked to a site datum.

As the Statnamic device is reliant on a remote settlement measuring system, it is advised that the reference source or measuring device is placed significantly far away from the Statnamic device to avoid the influence of the test on readings during the loading and unloading phase of the test. Recommendations for minimum separation requirements are given in table 2. In addition the separation of the reference/measuring device from the Statnamic rig should be measured for all tests and a note made of the ground surface that the device is placed.

In order to measure acceleration directly, allow for improved velocity calculations and verification of settlement measurements, it is recommended that the Statnamic loading device incorporates an accelerometer located as close as possible to the test pile. Ideally an additional accelerometer should be mounted on the pile. In order to allow accurate analysis of the Statnamic test all measurements must be logged at a rate in excess of 1 kHz and preferentially at rates of 4 kHz and above.

5   CONCLUSION

Where Statnamic tests are undertaken in clay soils, non-linear settlement dependant analysis should be undertaken with soil specific rate parameters. To aid this analysis Statnamic test termination criteria should be based upon achieved settlement and not the applied load. Maximum and preferably permanent settlements should be at least 10% of the pile's diameter. To aid specification of equipment and avoid under loading, a simple approach to estimating target load in clays would be to apply twice the anticipated ultimate static load. The accuracy of Statnamic analysis would benefit considerably from the inclusion of an accelerometer at the pile head and preferably with an additional unit attached to the pile. High-precision optical leveling of the pile before and after each loading event would allow multiple load cycles to be considered cumulatively and allow verification of settlement measurements. Prior to testing, the separation of the laser reference source from the test pile should be selected to avoid test induced surface wave disturbance during settlement measurement.

ACKNOWLEDGEMENT

Thanks go to my previous co-authors particularly Prof. Adrian Hyde and Dr John Powell. I would like to thank Peter Middendorp for his continuing technical assistance along with Stent Foundations Limited for access to case study data. The Grimsby research project was undertaken at the University of Sheffield, funded by the Engineering and Physical Research Council (EPSRC, UK, Grant GR/R46939/01) and supported by Expanded Piling Ltd, UK, Precision Monitoring and Control, UK and Berminghammer Foundation Equipment, Canada. Significant assistance was provided by Andrew Bell (Cementation Skanska Foundations Ltd) during this project. The current research project on rate effects in fine grained soils is funded by the Engineering and Physical Research Council (EPSRC, UK, Grant EP/E031749/1) and supported by BRE. The work will be undertaken at the University of Dundee.

REFERENCES

Balderas-Meca, J. (2004). Rate effects in rapid loading of clay soils. PhD Thesis, University of Sheffield, UK.

Brown, M.J. (2004). Rapid load testing of piles in fine grained soils. PhD Thesis, University of Sheffield, UK.

Brown, M.J., Anderson, W.F. and Hyde, A.F.L. (2004). Statnamic testing of model piles in a clay calibration chamber. Int. Journal of Physical Modelling in Geotechnics. Vol. 4, No. 1. pp. 11–24.

Brown, M.J. (2006). Static dynamite? – An introduction to Statnamic pile testing Proc. Swedish Geotechnical Society (Svenska Geotekniska Föreningen SGF) Grundlaggnings Dagen, Stockholm, Sweden, 9th March 2006. pp. 49–63.

Brown, M.J. and Hyde, A.F.L. (2006). Some observations of Statnamic pile testing. Proc. Inst. of Civil Engineers: Geotechnical Engineering Journal, Vol. 159, GE4. pp. 269–273.

Brown, M.J. & Hyde, A.F.L. (2008). Rate effects from pile shaft resistance measurements. Canadian Geotechnical Journal. Vol. 45, No. 3. pp. 425–431. DOI: 10.1139/T07-115.

Brown, M.J., Hyde, A.F.L. and Anderson, W.F. (2006). Analysis of a rapid load test on an instrumented bored pile in clay. Geotechnique. Vol. 56, No. 9, pp. 627–638.

Brown, M.J., Wood, T. and Suckling, T. (2006). Statnamic pile testing case studies. 10th Int. Conf. on Piling and Deep Foundations, Amsterdam, Holland, 31st May – 2nd June 2006, pp. 627–634.

Gudehus, G. (1981). Bodenmechanik. Institute of Soil Mechanics and Foundation Mechanics, University of Karlsruhe, Germany. (No longer in print, in German).

Institution of Civil Engineers (2007). Specification for piling and embedded retaining walls (SPERW). 2nd ed, Thomas Telford Publishing, London.

Janes, M.C., Justason, M.D. and Brown, D.A. (2000). Standard test for piles under rapid axial compressive load with its draft paper. Proc. 2nd Int. Statnamic Seminar, Tokyo, 28–30 October 1998, pp. 199–218.

Kusakabe, O. and Matsumoto, T. (1995). Statnamic tests of Shonan test program with review of signal interpretation. Proc. of the 1st Int. Statnamic Seminar, Vancouver, 27–30 September 1995, pp. 113–122.

Litkouhi, S. & Poskitt, T.J. (1980). Damping constants for pile driveability calculations. Geotechnique, Vol. 30, No. 1, pp. 77–86.

McVay, M.C., Kuo, C.L. and Guisinger, A.L. (2003). Calibrating resistance factor in load and resistance factor design of Statnamic load testing. Florida Dept. of Transportation, March 2003, Research Report 4910-4504-823-12.

Paikowsky, S.G. (2004). Innovative load testing systems. Geosciences Testing and Research Inc, Massachusetts, USA. National Cooperative Highway Research Program, September 2004, Research Report NCHRP 21-08.

Powell, J.J.M. and Brown, M.J. (2006). Statnamic pile testing for foundation re-use. In A.P. Butcher, J.J.M. Powell & H.D. Skinner (eds) Int. Conf. on the Re-use of Foundations for Urban Sites, Watford, UK, 19–20th October 2006, pp. 223–236.

Powell, J.J.M. and Quaterman, R.S.T. (1988). The interpretation of cone penetration in clays with particular reference to rate effects. Proc. Int. Symp. On Penetration Testing, Orlando, pp. 903–910.

Randolph, M.F. and Deeks, A.J. (1992). Dynamic and static soil models for axial pile response. Proc. 4th Int. Conf. on the Application of Stresswave Theory to Piles, The Hague, September 1992, pp. 3–14.

The Japanese Geotechnical Society (2000). Draft of method for rapid load testing of single piles (JGS 1815-2000). Proc. 2nd Int. Statnamic Seminar, Tokyo, 28–30 October 1998, pp. 237–242.

Chapter 3

Load transfer in rapid load pile tests in clays

Adrian F.L. Hyde
University of Sheffield, UK

Michael J. Brown
University of Dundee, UK

SUMMARY

In order to gain an insight into the behaviour of rapid load pile testing in clays, a full scale pile instrumented with accelerometers, strain gauged sister bars and a tip load cell was tested in a glacial lodgement till near Grimsby, UK. The soil around the pile was also instrumented with radially arrayed buried accelerometers. The test pile was subjected to rapid loading tests, the results of which were compared with constant rate of penetration and maintained load static tests on the same pile. Shaft frictions derived from the strain gauged reinforcement in the pile have been compared with shear strains and stresses derived from accelerations in the surrounding soil to give an insight to the load transfer mechanisms for a rapidly loaded pile in clay.

1 INTRODUCTION

Traditionally, pile load testing is carried out using static methods which are relatively simple to undertake and analyse but require substantial temporary infrastructure that increases with pile ultimate load capacity making them expensive and time consuming. Rapid load test (RLT) methods have been developed as an alternative which utilise the advantages of both static and dynamic tests while avoiding some of the disadvantages. In order to gain an insight into the behaviour of rapid load pile testing in clays, a full scale pile instrumented with accelerometers, strain gauged sister bars and a tip load cell was tested in a glacial lodgement till near Grimsby, UK.

2 GENERAL REVIEW OF THE RESEARCH

In recent decades dynamic pile testing methods have been developed which require minimal infrastructure but are complicated to analyse. They may also result in pile damage due to the generation of tensile stresses as a result of the very short loading period of 5 to 10 milliseconds. An alternative rapid load testing method, known commercially as the Statnamic test, has been developed with a longer loading period. It works by the rapid burning of a fuel that produces gas in a pressure chamber (Middendorp 1993). This gas accelerates a reaction mass upwards at a maximum peak acceleration of about 20 g that in turn imparts a downward load on the test pile.

Thus, only 5% of the reaction mass used during static testing is required to produce the same load (Middendorp 2000). The maximum load and its duration are regulated by controlling the quantity of fuel and the venting of the gas. The load duration is normally regulated to about 200 milliseconds. This loading period is between 20 and 40 times greater than that for a dynamic test, thus avoiding the generation of tensile stresses for piles of up to 40 m length (Nishimura & Matsumoto 1995, Middendorp & Bielefeld 1995, Mullins *et al.* 2002).

For foundation design, it is necessary to derive an equivalent static load-settlement curve from the rapid load test data by eliminating rate effects. The most common form of analysis currently used is the unloading point method or UPM (Kusakabe & Matsumoto 1995) which takes into account both velocity dependant soil viscous damping and acceleration dependant pile inertia. However, this method assumes the soil viscous damping is linear with velocity. Soil damping in clays is highly nonlinear and forms an important component of ultimate pile resistance at enhanced rates of testing (Hyde *et al.* 2000, Brown 2004). The UPM method provides a good correlation with static tests for sands and gravels (Brown 1994, McVay *et al.* 2003, Wood 2003) where viscous damping is negligible, but over predicts pile capacities by up to 50% for clay soils (Holeyman *et al.* 2000).

3   DESCRIPTION AND RESULTS OF THE RESEARCH

3.1   Full scale pile test

In order to gain a better understanding of the load transfer mechanisms under rapid loading and to improve the analysis of these types of tests in clay soils, a full scale auger bored pile, instrumented with strain gauges and accelerometers, was installed in a glacial lodgement with buried accelerometers. Rapid load test results are compared with those from standardised static load tests.

The ground conditions at the test site comprised matrix dominant glacial lodgement till (Weltman & Healy 1978), which was underlain by Cretaceous chalk bedrock (Powell & Butcher 2003). The till of this region is described as being stiff to firm, greyish to dark brown, predominantly silty clay with a variety of cobbles boulders and pebbles (Berridge & Pattison 1994). It is cohesive, overconsolidated, but may also be soft and weathered (reddish brown) with grey joint surfaces (Bell 2001).

The instrumented test pile was a 600 mm nominal diameter bored cast insitu pile 12.0 m in length. A steel friction reducing casing was inserted to 1.8 m BGL to reduce the influence of the near surface variation in soil properties. The reinforcement consisted of six vertical 12 m long T16 bars with a single T12 helical at 300 mm vertical centres. The pile concrete had a 28 day strength of 36 N/mm^2 and an average density of 2.35 Mg/m^3.

Fifteen 1 m long strain gauged sister bars were incorporated in the pile, three at each of five different levels, tied to the horizontal reinforcement (Fig. 1). The three bars at each level were spaced equally around the pile circumference. Two piezoelectric shear type accelerometers were also fixed to the pile reinforcement at two levels.

Six accelerometers were also installed in the soil surrounding the pile. As shown in figure 1 these were installed at two levels, 4 m & 8 m below ground level (BGL)

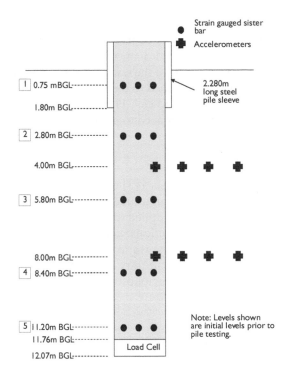

Figure 1 Pile instrumentation.

corresponding to the levels of the accelerometers installed in the pile, with three at each level (4 m BGL: 3.05R, 5.04R & 8.5R; 8 m BGL: 2.89R, 4.70R & 9.12R, where R was the pile radius). The radial locations of the accelerometers were chosen based upon accelerations measured in a related laboratory calibration chamber study into the behaviour of model piles subject to rapid loading (Brown 2004). To install the accelerometers a hollow casing of 36 mm internal diameter with a sacrificial tip was pushed to the required depth. The accelerometer in its own protective casing was lowered down the hollow casing until contact was made with the tip. The casing was then withdrawn as a cement/bentonite grout was poured down the casing.

The pile testing programme commenced 35 days after the pile was installed and was designed to compare the results from rapid and static load tests. The pile was subjected to five different target levels of rapid loading over a two day period with an initial test taken to 1163 kN on the first day followed by peak loads of 1700, 2048, 2525 and 3071 kN on the second day. The actual peak loads were always higher than the related target loads. This was followed 21 days later by a constant rate of penetration (CRP) test carried out at 0.01 mm/s and a further 5 days after this by a maintained load test (MLT). The maintained load test was undertaken as a proof load test followed by an extended proof load test to the same specification. The design verification load chosen was 900 kN with a specified working load of 900 kN (ICE 1997).

For rapid loading, a 3 MN tripod Statnamic device was used with a hydraulic catch mechanism and 18 tonne weight pack. Load was measured by a load cell mounted in the base of the piston device. The pile settlement was measured by a photovoltaic sensor mounted in the base of the piston which was excited by a laser reference beam (Middendorp 2000). For static loading, reaction was provided by beams running over the test pile anchored to three 11.5 m long, 600 mm auger bored diameter piles arranged in a triangular pattern with two piles equidistant at 2.1 m from the test pile (centre to centre) and a third pile at the apex, 3.5 m distant. Load was applied to the test pile by a hydraulic jack and measured by an independent calibrated load cell. Pile settlement readings were provided by 4 LVDTs placed on the pile and referenced to a remote beam.

3.2   Load and settlement behaviour

The load-settlement behaviour for each of the pile tests is shown in figure 2. It is immediately apparent from comparison of the higher magnitude rapid load tests with the static tests that, although much greater loads were applied to the pile during rapid load testing, the resulting maximum and residual deflections were smaller. Pile yield and ultimate resistance can be easily identified from the static test data. This is not so evident on inspection of the results from the rapid load tests despite the maximum load for the 3000 kN test exceeding the maximum static loads for CRP and MLT by a factor of 1.39 and 1.71 respectively.

4   SHAFT CAPACITY AND LOAD TRANSFER

To analyse the results from the embedded sister bars the stiffness of the pile was calculated from the results of the MLT and used for the load transfer analyses for the CRP and rapid load tests (Brown *et al.* 2006). The results from the load transfer

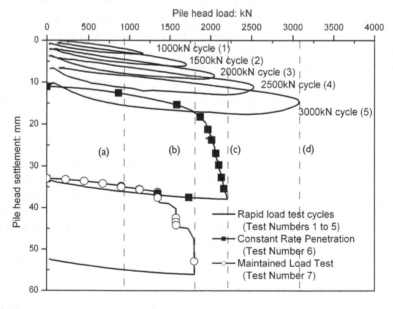

Figure 2 Pile load tests.

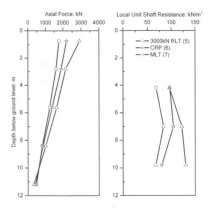

Figure 3 Axial loads and shaft resistance at maximum (b), (c), (d).

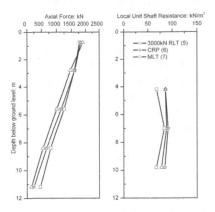

Figure 4 Axial loads and shaft resistance at 1800 kN (b).

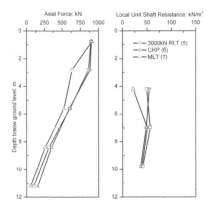

Figure 5 Axial loads and shaft resistance at 900 kN (a).

analysis for the rapid load test with a target load of 3000 kN are shown compared with the CRP and MLT at various loads in figures 3 to 5.

The load transfer from pile to soil was calculated from the change in pile load measured between two points. Examination of the change in axial force between the upper strain gauge positions shown in figures 3 to 5 reveals that considerably greater load transfer occurred during rapid load testing than during the other types of test. This greater load transfer particularly between Levels 1 and 2 for rapid load tests could be due the upper very stiff clay layer. Where axial loads are compared at 900 & 1800 kN (Figs. 4 & 5), loads in the pile shaft during rapid load testing are slightly lower than those during static testing, indicating greater load transfer had occurred at the top of the pile.

In figure 3 the maximum load levels are different in each case. However, the decay in axial force in the pile is greater for the rapid load test indicating a higher degree of load transfer. This said, it is difficult to establish if similar conditions of pile mobilisation had been reached in the rapid load test as in the static tests. Although far greater loads were applied during the rapid load test, much smaller residual deflections were observed.

The shaft resistances shown in figure 3 illustrate the enhancement associated with rapid load tests at ultimate load conditions. The shaft resistance between 4 m and 10 m BGL for the rapid load test varied from 96 to 130 kN/m^2 at peak load which is approximately 30% higher than the MLT results and 20% greater than the CRP values. At peak load during the rapid load test, the pile head velocity was 475 mm/s compared to 0.01 mm/s during the CRP test. This confirms that pile shaft resistance is significantly influenced by pile loading rate after soil yielding has occurred. In contrast, the shaft resistances below Levels 1 and 2 for the rapid load test at load levels of 900 & 1800 kN are actually similar or lower than those encountered in the static load tests even though the pile head velocities were 70 and 290 mm/s respectively. At these load levels the 3000 kN rapid load test result was still in the pre-yield phase rather than post yield plastic phase of loading and, much of the load transfer was occurring in the stiff near surface clay layer. When the rapid load test results are compared with the static results in the pre-yield linear phase of the test there is no significant enhancement of stiffness again implying that the rate dependent or viscous behaviour was not initiated until yield of the soil had occurred.

5   SOIL BEHAVIOUR DERIVED FROM GROUND ACCELERATIONS

Typical vertical accelerations measured in the pile and the surrounding soil at 4 m BGL can be seen in figure 6. Figure 7 shows that the ground accelerations had decayed markedly by a radial distance 3R and had fallen below 20% of the pile's acceleration by 6R. The changes in the ratio of soil to pile acceleration with radial distance from the pile at different load levels during the 3000 kN rapid load cycle are shown in figures 8 & 9 for transducers located at 4 m and 8 m BGL respectively. It is clear that the ground accelerations vary with changing load level during a rapid load test. These accelerations are at a maximum at low loads reducing to a minimum at peak load. This would suggest a decoupling of the pile acceleration from that of the soil on approaching peak load as originally proposed by Randolph and Simons (1986) who suggested that acceleration of the soil mass would only occur during the initial pre-yield linear loading phase. After this, the pile and a relatively small annulus of soil would shear relative to the larger soil mass and rate dependant soil shear resistance

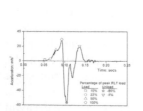
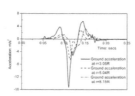
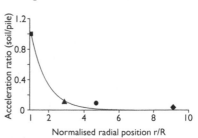

Figure 6 Pile and ground accelerations 4 m BGL, 3000 kN RLT.

Figure 7 Normalised radial acceleration.

would need to be considered. This model of pile-soil load transfer is consistent with the measurements of shaft resistance discussed above where little enhancement of shaft resistance was noted prior to yielding. This said, the soil again begins to accelerate with the pile during unloading. For instance, in figure 8 during unloading the proportion of acceleration experienced by the soil at 80% peak RLT load is similar to that observed at 50% peak RLT load during the loading phase.

The average shear strain in the soil increased with increasing applied pile load up to peak load (Figs. 10 & 11). After this the shear strain continued to increase as the pile unloaded. The results show that during the RLT test the shear strains at 2R and beyond may be classified as small (as defined by Viggiani & Atkinson, 1995) with large strains only occurring at 2R at maximum pile displacement and beyond.

The radial variation of average vertical shear stress normalised by the undrained shear strength is shown in figures 12 & 13. The shear stresses are consistent with the strains except that the stresses at the pile-soil interface obtained from the pile instrumentation are a maximum at the peak applied pile load. During unloading the shaft resistance reduced rapidly resulting in a lag between the change in soil stress

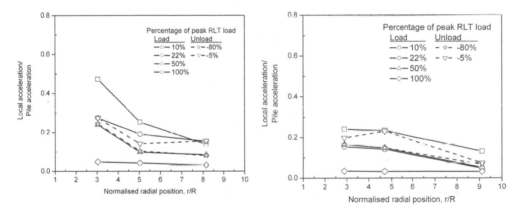

Figure 8 Normalised soil acceleration 4 m BGL.   *Figure 9* Normalised soil acceleration 8 m BGL.

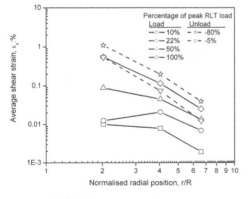
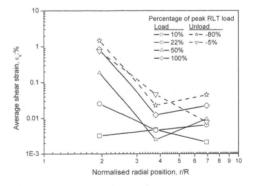

Figure 10 Soil shear strain 4 m BGL.   *Figure 11* Soil shear strain 8 m BGL.

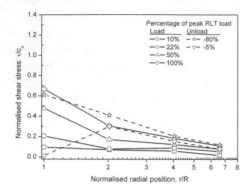

Figure 12 Soil shear stress 4 m BGL.

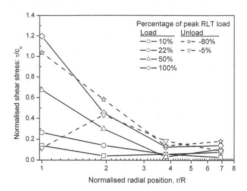

Figure 13 Soil shear stress 8 m BGL.

close to the pile and that at distance such that when the load reduced to 5% of the peak value the shear stress at 2R was higher than the shaft friction.

In Figs. 8 & 9 the acceleration ratios from 4.5R outwards were similar for the two levels of 4 m and 8 m, the proportion of acceleration at 3.05R at 4 m BGL was greater for all but the peak load. This is likely to have been influenced by the degree of mobilisation of soil shear strength which as can be seen in figures 12 & 13 was significantly lower at 4 m BGL than at 8 m BGL.

Consequences of the results for the Standard and the guidelines.

* Viscous rate effects are not consistent throughout a rapid load test in clay and hence algorithms correcting for these effects should allow for enhanced viscosity after yielding of the pile soil interface.
* Damping effects can be considered to be small up to 50% of the ultimate static pile capacity in clays.
* Initial load displacement properties under static and rapid loading are similar in clays.

6   CONCLUSIONS

1   Enhanced shaft resistance due to viscous rate effects in clays occurs predominantly after yield of the pile soil interface.
2   Low ground accelerations at peak load in rapid load testing indicated constant velocity decoupling.
3   Soil shear stresses and shear strains were maximum at peak load indicating maximum energy dissipation.
4   Soil accelerations reduced t o <1% at >2R.

ACKNOWLEDGEMENTS

This research project was funded by the Engineering and Physical Sciences Research Council (Grant No. GR/R46939/01). Technical assistance was provided by

Berminghammer Foundation Equipment (Canada) and TNO Building Research (Netherlands). The Expanded Piling Company (UK) kindly installed the piles and provided the research site facility. Rapid load testing and static pile load testing was undertaken by Precision Monitoring and Control (UK).

REFERENCES

Bell, A. (2001). Investigation into the increase in capacity with time of precast piles driven into stiff overconsolidated clay. MSc Thesis, University of Sheffield, UK.

Berridge, N.G. and Pattison, J. (1994). Geology of the country around Grimsby and Partington. British Geological Survey. 1983, Memoir.

Brown, D.A. (1994). Evaluation of static capacity of deep foundations from Statnamic testing. Geotechnical Testing Journal, ASTM, Vol. 17, No. 4, pp. 403–414.

Brown, M.J. (2004). The rapid load testing of piles in fine grained soils. PhD Thesis, University of Sheffield, UK.

Brown M.J., Hyde A.F.L. and Anderson W.F. (2006). Analysis of a rapid load pile test on an instrumented bored pile in clay, Geotechnique Vol. 56. No. 9, pp. 627–638.

Holeyman, A.E., Maertens, J., Huybrechts, N. and Legrand, C. (2000). Results of an international pile dynamic testing prediction event. Proc. 6th Int. Conf. on the Application of Stress Wave Theory to Piles, Sao Paulo, pp. 725–732.

Hyde, A.F.L., Robinson, S.A. and Anderson, W.F. (2000). Rate effects in clay soils and their relevance to Statnamic pile testing. Proc. 2nd Int. Statnamic Seminar, Tokyo, pp. 303–309.

Institution of Civil Engineers. (1997). Specification for piling and embedded retaining walls. 1st ed. London, UK, Thomas Telford Publishing.

Kusakabe, O. and Matsumoto, T. (1995). Statnamic tests of Shonan test program with review of signal interpretation. Proc. 1st Int. Statnamic Seminar, Vancouver, pp. 113–122.

McVay, M.C., Kuo, C.L. and Guisinger, A.L. 2003. Calibrating resistance factor in the load and resistance factor design of Statnamic load testing. Research Report 4910-4504-823-12, Contract BC354, RPWO#42. University of Florida, Florida Department of Transportation.

Middendorp, P. (1993). First experiences with Statnamic load testing of foundation piles in Europe. Proc. 2nd, Int. Geotechnical Seminar on Deep Foundations on Bored and Auger Piles, Ghent, pp. 265–272.

Middendorp, P. (2000). Statnamic the engineering of art. Proc. 6th Int. Conf. on the Application of Stress Wave Theory to Piles, Sao Paulo, pp. 551–561.

Middendorp, P. and Bielefeld, M.W. (1995). Statnamic load testing and the influence of stress wave phenomena. Proc. 1st Int. Statnamic Seminar, Vancouver, pp. 207–220.

Mullins, G., Lewis, C.L. and Justason, M.D. (2002). Advancements in Statnamic data regression techniques. Proc. ASTM Conf. Int. Deep Foundations Congress, Florida, ASTM Geotechnical Special Publication No.116, Vol. 2, pp. 915–930.

Nishimura, S. and Matsumoto, T. (1995). Wave propagation analysis during Statnamic loading of a steel pipe. Proc. of 1st Int. Statnamic Seminar, Vancouver, pp. 23–33.

Powell, J.J.M. and Butcher, A.P. (2003). Characterisation of a glacial till at Cowden, Humberside. Proc. Int. Conf. on the Characterisation and Engineering Properties of Natural Soils, Singapore, pp. 983–1019.

Viggiani, G. and Atkinson, J.H. (1995). Stiffness of fine-grained soil at very small strains. Geotechnique, Vol. 45, No. 2, pp. 249–265.

Weltman, A.J. and Healy, P.R. (1978). Piling in Boulder Clay and other glacial tills. CIRIA Report No. PG5, DOE & CIRIA Piling Group, London.

Wood, T. (2003). An investigation into the validation of pile performance using Statnamic tests. MSc Thesis, Imperial College, UK.

Chapter 4

Practice of rapid load testing in Japan

T. Matsumoto
Department of Civil Eng., Kanazawa University, Kakuma-machi, Kanazawa, Japan

SUMMARY

This paper reviews the research activities for standardisation of rapid load test of piles which stated from the early of 1990s, and presents developments and current practices of rapid load tests of piles in Japan.

I INTRODUCTION

The rapid load test method first used in Japan was the Statnamic test developed by Bermingham & Janes (1989). The first Statnamic test in Japan was performed in 1991 by Takenaka Corporation against a cast-in-place concrete pile. In 1992, comparable tests of static and the Statnamic load tests were carried out for a steel pipe pile (Matsumoto *et al.* 1994). The results and performance of these Statnamic tests demonstrated a promising potential of rapid load test for an alternative to the conventional static load test. Hence, in 1993, a Research Group of Rapid Pile Load Test Methods (the Research Group hereafter) was established led by Prof. Osamu Kusakabe, Tokyo Institute of Technology.

2 RESEARCH ACTIVITIES ON RAPID LOAD TEST

The primary objectives of the Research Group were

1   To compile the existing knowledge about rapid load test,
2   To examine basic characteristics and applicability of the rapid load test and
3   To establish scientific interpretations of the rapid load test results (Kusakabe 1998).

A number of comparative tests of static load test and Statnamic test were carried out in the activity of the research group. The achievements of the research group were presented in the Proceedings of the 1st and the 2nd International Statnamic Seminars in 1995 and 1998.

Owing to the research activity of the research group, the Research Committee on Rapid Pile Load Test Methods was established in the Japanese Geotechnical Society in 1996 and it was upgraded to the Standardization Committee on Vertical Load Test Methods in 1998. Through these activities, rapid pile load test method was implemented in Standards for Vertical Load Tests of Piles (Japanese Geotechnical Society 2002), and the research group served their purpose in 2002.

3   JAPANESE STANDARD FOR RAPID LOAD TEST OF SINGLE PILES

In the Standards of Japanese Geotechnical Society for Vertical Load Tests of Piles, rapid load test is categorised into a kind of dynamic method. Distinguish between rapid load test and dynamic load test is made by means of the relative loading duration, T_r, that is defined by

$$T_r = t_L / (2L/c) \tag{1}$$

where t_L is the loading duration, L is the pile length and c is the bar wave velocity.

Pile head loading with $T_r \geq 5$ is regarded as rapid load test in which wave propagation phenomena can be neglected in the analysis or interpretation of the test signals, according to Nishimura et al. (1998). Note that it is premised on that the pile head load increases with time smoothly to the peak and then smoothly decreases without any oscillation. It is prescribed in the standard that a single mass model may be used to derive static load displacement relationship from the test signals in case of rapid load test, and that other approaches such as one-dimensional stress wave analysis and finite element analysis may also be used to examine the dynamic or static pile behaviour observed during the test.

In the standard, pile load test methods are classified in terms of the influence of stress-wave propagation in pile alone. However, stress-wave propagation phenomena occur also in the ground. Shear stress-wave propagation radiating from the pile shaft is predominant in the ground surrounding a friction pile. Delay of the ground deformation (mainly shear deformation) behind the stress change in the soil occurs due to the wave propagation phenomena. Such dynamic behaviour of soil deformation has been modelled by Novak et al. (1978) as shown in figure 1, assuming the deformation of elastic soil under plane strain condition.

The values of spring, k_s, and radiation damping, c_r, are given as follows approximately for loading of relatively high frequencies (Novak et al. 1978):

$$k_s = 2.75 G / (\pi d), \quad c_r = G / V_s \tag{2}$$

where G and V_s are the shear modulus and the shear wave velocity of the soil, and d is the pile diameter.

The author has proposed a measure, R_T, for loading duration where the radiation damping becomes virtually negligible, based on the theoretical model proposed by Novak et al. (1978) (Research Committee on Rapid Pile Load Test Methods 1998). If the shear stress f is applied abruptly and maintained for a long time as shown in figure 2, the force equilibrium of the soil system in figure 1 is expressed as:

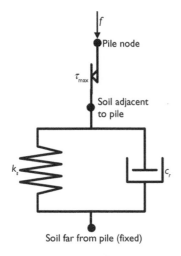

Figure 1 Model of soil surrounding
pile under dynamic loading.

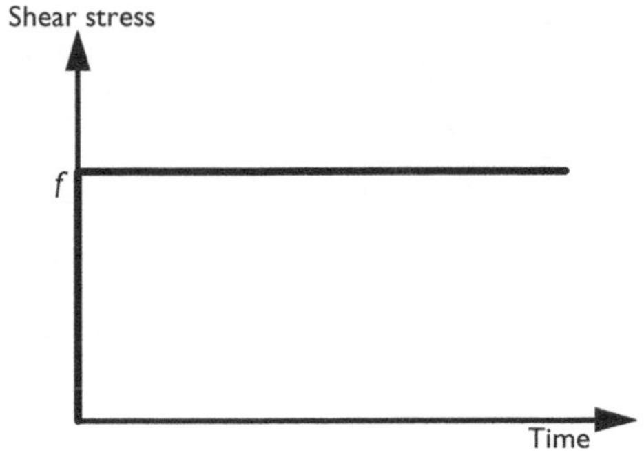

Figure 2 Change of shear stress, f, with time.

$$f = k_s w_s + c_r \dot{w}_s \qquad (3)$$

where w_s is the vertical displacement of the soil adjacent to the pile shaft.

The solution of Eq. (3) is given as follows with the initial conditions that $w_s = \dot{w}_s = 0$ at time $t = 0$.

$$w_s(t) = \frac{f}{k_s}\left[1 - \exp\left(-\frac{k_s}{c_s}t\right)\right] \qquad (4)$$

The ratio k_s/c_s is approximated as follows from Eq. (2):

$$\frac{k_s}{c_r} = \frac{2.75 V_s}{\pi d} \approx \frac{V_s}{d} \qquad (5)$$

Hence, Eq. (4) is rewritten as

$$w_s(t) = \frac{f}{k_s}\left[1 - \exp\left(-\frac{V_s}{d}t\right)\right] \qquad (6)$$

The change with time of soil displacement w_s is shown in figure 3. In figure 3, the soil displacement is normalized by f/k_s (final soil displacement) and time is normalized by d/V_s so that $R_T = t/(d/V_s)$. R_T is called 'relative relaxation time'. For an example, if difference of soil displacements in static loading and rapid (or dynamic) loading is prescribed to be 5% [$w_s(t) = 0.95 w_s(\infty)$], the corresponding relative relaxation time R_T is 3. For piles having 1 m diameter in soils having $V_s = 300 - 30$ m/s, relaxation time,

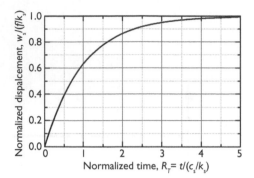

Figure 3 w_s vs t in normalized expression.

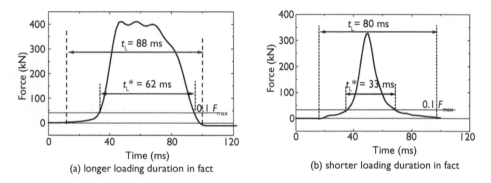

Figure 4 Different loading patterns having similar loading duration.

$t_R = R_T \times (d/V_s)$, ranges from 0.01 s (10 ms) to 0.1 s (100 ms). Loading duration of rapid load testing must be longer than the 'relaxation time' to eliminate effects of the radiation damping of the shaft resistance.

Another issue to be considered is the definition of the loading duration t_L. Figure 4 shows examples of loading patters on a pile by different loading methods. If we define the loading duration as time duration from the start to the end of loading, the loading duration of figure 4(a) is 88 ms that is comparable to the loading duration of 80 ms in figure 4(b). However, comparing both figures, practical loading duration of figure 4(b) is much shorter than that of figure 4(a) intuitively. If effective loading duration, t_L^*, is defined as loading duration over $0.1F_{max}$ where F_{max} is the maximum of the applied force, t_L^* are 62 ms and 33 ms for figure 4(a) and figure 4(b), respectively. This difference of t_L^* seems to express the difference of practical loading durations of figure 4(a) and figure 4(b). It is recommended to describe how to define the loading duration in standard or guideline.

One of advantages of the rapid load test (Statnamic test) is that the load is measured via a calibrated load cell and does not rely on pile material and cross-section properties

(Poulos 1998). Most piles in Japan are cast-in-place concrete piles or pre-fabricated concrete piles using auger construction method. Measurement of the applied load via a load cell is preferable for these pile types. The load shall be measured via a load cell in rapid load test, in the author's opinion.

4 DEVELOPMENTS OF RAPID LOAD TEST METHODS

More than 100 Statnamic tests had been carried out until 2001 in Japan. However, the number of the Statnamic tests did not increase so much after the force of the standard, because time and cost of the Statnamic test. It takes usually 2 days to perform a Statnamic test including preparation, test and disassembly of the test equipment. It is also difficult to perform cyclic loading using the Statnamic.

Other rapid pile load test methods have been developed, e.g. Dynatest by Gonin and Leonard (1984) and Pseudo-static pile load tester by Schellingerhout & Revoort (1996). These rapid pile load test methods utilise a falling mass and coil springs on the pile top to prolong loading duration. The falling mass type rapid load test methods offer advantages over the Statnamic, including:

i   time for preparation of test equipment for a pile is very short, less than 10 min in usual.
ii   it is very easy to conduct cyclic or repetitive loading for a pile.
iii   loading duration is easily adjusted by changing combination of mass of the hammer and stiffness of the spring.
iv   maximum load is easily controlled by changing the falling height of the hammer mass.
v   cost of the test may be cheaper than that of the Statnamic.

Falling mass type rapid load test methods have been developed also in Japan: Soft Cushion Method by System Keisoku Corporation, Spring Hammer test method by the joint of Kanazawa University and Marubeni Material Lease Corporation (Matsumoto *et al.* 2004) and Hybridnamic Load Test Method by JibanShikenjo Corporation.

Figure 5 Soft cushion rapid load test method by System Keisoku Corporation.

Figure 6 The Spring Hammer test device mounted on a pile construction machine.

Loading principle of these test methods is basically identical to that of Dynatest and Pseudo static test methods, although some devisal are made for spring or cushion on the pile top. Figure 5 shows the Soft Cushion Test device having a loading capacity to measure a static capacity of about 16 MN. Figure 6 shows the Spring Hammer loading device having a loading capacity of 2 MN. The hammer and the spring unit are mounted on a leader of a pile construction machine together with a screw auger. An advantage of this test device is that rapid pile load tests can be done promptly after the construction of an auger pile. Of course, the Spring Hammer device can be used for rapid load test on any constructed pile.

One of advantages of the rapid load test is that simplified interpretation methods, in which the pile is treated as a rigid mass neglecting wave propagation phenomena in the pile, could be used to derive a static load-settlement relation from the measured dynamic signals.

Figure 7 shows the modelling of pile and soil during rapid pile load testing. The pile is assumed as a rigid mass having mass of M, and the soil is modelled by a spring and a dashpot. This modelling is advocated by Middendorp *et al.* (1992) and Kusakabe and Matsumoto (1995). The Unloading Point Method (Kusakabe and Matsumoto 1995) is widely used in the interpretation of rapid load test signals in Japan, in which the damping value C is assumed to be linear while soil spring is treated as non-linear. Within the author's experiences, the Unloading Point method tends to overestimate the pile head stiffness of the test pile in many cases.

Matsumoto *et al.* (1994) have proposed a simplified interpretation method of rapid load test signals, in which both of the soil spring K, and the damping value C are assumed to be non-linear. This method is called 'non-linear damping method' (note that the method was called 'modified initial stiffness method' in the original paper).

The non-linear damping method to obtain the 'static' load-settlement relation' of a test pile is as follows:

Figure 8 shows the notations used in the non-linear damping method. The applied load F_{rapid} is equal to the sum of the soil resistance F_{soil}, and the inertia of the pile:

$$F_{soil}(i) = F_{rapid}(i) - M \cdot \alpha(i) \tag{7}$$

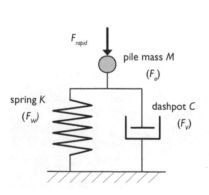

Figure 7 Modelling of pile and soil during rapid loading.

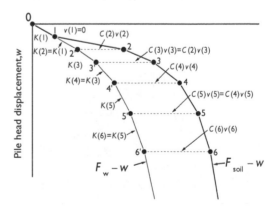

Figure 8 Notations used in non-linear damping interpretation method.

where M is the pile mass and $\alpha(i)$ is the measured pile acceleration at time step i.

The soil resistance F_{soil} is the sum of the spring resistance (static resistance) F_{w}, and the dashpot resistance, F_{ν}.

$$F_{\text{soil}}(i) = F_{\text{w}}(i) + F_{\nu}(i) = F_{\text{w}}(i) + C(i) \cdot v(i) \tag{8}$$

where $C(i)$ is the damping coefficient and $v(i)$ is the pile velocity at time step i.

At the first step ($i = 1$), the initial stiffness, $K(1)$, is calculated by the initial static load $F_w(1)$ divided by the initial displacement $w(1)$.

$$K(1) = F_{\text{W}}(1) / w(1) = F_{\text{static}} / w_{\text{static}} \tag{9}$$

At the next step (at step $i + 1$), the soil spring $K(i + 1)$ is assumed to be equal to $K(i)$ as indicated by Eq. (10). Hence, the static resistance $F_w(i + 1)$ at step $i + 1$ is calculated by Eq. (11). The value of $C(i + 1)$ can be determined by means of Eq. (12).

$$K(i + 1) = K(i) \tag{10}$$

$$F_{\text{w}}(i + 1) = F_{\text{w}}(i) + K(i + 1) \cdot \{w(i + 1) - w(i)\} \tag{11}$$

$$C(i + 1) = \{F_{\text{soil}}(i + 1) - F_w(i + 1)\}/v(i + 1) \tag{12}$$

At the following step $i + 2$, $C(i + 2)$ is assumed to be equal to $C(i + 1)$ as indicated by Eq. (13). Therefore, the values of $F_w(i + 2)$ and $K(i + 2)$ can be determined by means of Eqs. (14) and (15), respectively.

$$C(i + 2) = C(i + 1) \tag{13}$$

$$F_w(i + 2) = F_{\text{soil}}(i + 2) - C(i + 2) \cdot v(i + 2) \tag{14}$$

$$K(i + 2) = \frac{F_w(i + 2) - F_w(i + 1)}{w(i + 2) - w(i + 1)} \tag{15}$$

By repeating the procedure from Eq. (10) to Eq. (15), the values of K and C for following steps are alternately updated consecutively. Finally, whole static load-displacement relation F_w vs. w is constructed as shown in Figure 8.

5 A CASE OF COMPARATIVE STATIC AND RAPID LOAD TESTS

Comparisons of static load tests and rapid load tests were carried out to validate the rapid load testing using the Spring Hammer device (Matsuzawa *et al.* 2006, Nakashima *et al.* 2006). The results of these comparative tests are briefly introduced.

Figure 9 shows the profiles of the soil layers and SPT N-values at the test site. The ground at the test site blow a depth of 2 m consists of alternating layers of silt and sand. The SPT N-values are less than 7 to a depth of 7 m, and about 15 from 8 m to 11 m in depth. A total of 7 test piles were constructed. Static and rapid load tests were performed on pile No.1 and pile No. 3 out of the 7 test piles (see Table 1).

Each test pile was steel H-shaped pile with an end circular plate so as to increase the end-bearing capacity, as shown in Figure 10. Construction procedure of the test pile is as follows:

i   borehole is made by a screw auger.
ii   fresh cement mortar is poured in the borehole.
iii   pile is inserted into the borehole.
iv   the inserted pile is driven by a drop hammer. During this stage, rapid pile load test is carried out using the Spring Hammer device.
v   when the required end-bearing capacity is obtained, driving and testing are terminated.

One of advantages of this pile and construction procedure is that performance of all the constructed piles can be obtained promptly at the site. The pile is called MSSP (Marubeni Super Safety Pile).

Figure 11 shows the measured dynamic signals during rapid load test of pile No. 3 during construction. The force on the pile top is measured by a load cell, the displacement and accelerations at the pile head are measured by an optical displacement transducer and accelerometers. In this test, load duration is 0.14 s (140 ms) that can

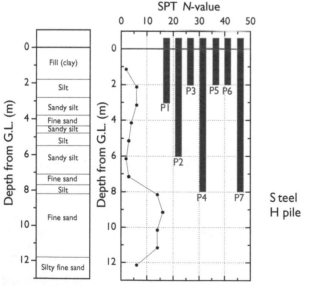

Figure 9 Profiles of soil layers and SPT N-values, together with seating of the test piles.

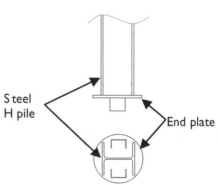

Figure 10 Steel H pile with end-plate.

Table 1 Specifications of piles 1, 3 and 7.

	Pile 1	Pile 3	Pile 7
Length (m)	4.0	3.0	9.0
Embedment length (m)	3.0	2.0	8.0
Diameter of borehole (m)	0.45	0.45	0.55
Diameter of end plate (m)	0.45	0.45	0.47
Load test type at construction	Rapid	Rapid	Rapid
Load test type 7 days after the end of construction	Static and Rapid	Static and Rapid	None

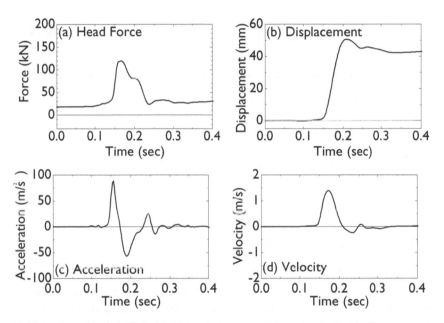

Figure 11 Examples of measured dynamic signals (construction stage of pile No. 3).

be regarded as rapid loading according to the *Standards of JGS for Vertical Load Tests of Piles*.

Static load tests of piles No. 1 and No. 3 were carried out 7 days after the end of construction. Note that bearing capacity of the test piles increased during this rest period, because the fresh cement mortar hardened during this period. Rapid load tests were carried out soon after the end of static load tests. The non-linear damping method was used to derive static load-settlement curve of the test pile for each blow.

Figure 12 shows the comparison of load-settlement curves from the static load test and the rapid load tests on pile No. 1. Three rapid load tests were carried out after the static load test by falling the hammer mass of 2 tons from heights, *h*, of 0.1, 1.0 and 1.5 m. It can be seen from figure 12 that the envelop curve of the curves from the static and rapid load tests is consistent.

Figure 13 shows the comparison of load-settlement curves from the static load test and the rapid load tests on pile No. 3. A consistent envelop of the curves from the

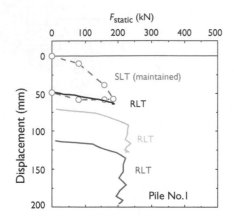

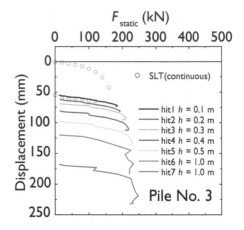

Figure 12 Comparison of results of static and rapid load tests on pile No. 1 conducted at 7 days after the end of pile construction.

Figure 13 Comparison of results of static and rapid load tests on pile No. 3 conducted at 7 days after the end of pile construction.

static and rapid load tests can be seen again. The results of figsures 12 and 13 encourage the use of the rapid pile load test with the non-linear damping interpretation method as an alternative to the static load test.

6   CONCLUDING REMARKS

Research, practice and process of standardization of rapid pile load testing from 1991 in Japan were briefly reviewed. Japanese standard for rapid pile load test was briefly introduced, and definition of rapid load test was discussed. Recent developments of rapid load test methods in Japan were also presented. Finally, a case study of comparisons of static and rapid load tests was presented to demonstrate that rapid load testing with an appropriate interpretation method can be used as an alternative to the static load test.

REFERENCES

Bermingham, P. and Janes, M. (1989). An innovative approach to load testing of high capacity piles. Proc. of the Int. Conf. on Piling and Deep Foundations, London: 409–413.

Gonin H.G.C. and Leonard M.S.M. (1984). Theory and performance of a new dynamic method of pile testing. Proc. of 2nd Int. Conf. of Application of Stress-Wave Theory to Piles, Stockholm: 403–410.

Japanese Geotechnical Society, (2002). Standards of Japanese Geotechnical Society for Vertical Load Tests of Piles. Japanese Geotechnical Society, Tokyo.

Kusakabe, O. (1998). Changing foundation design code and the role of Statnamic test. Proc. 2nd Int. Statnamic Seminar, Tokyo, Japan: 23–39

Kusakabe, O. and Matsumoto, T. (1995). Statnamic tests of Shonan test program with review of signal interpretation, Proc. 1st Int. Statnamic Seminar, Vancouver, Canada: 13–122.

Matsumoto, T., Tsuzuki, M. and Michi, Y. (1994). Comparative study of static loading test and Statnamic on a steel pipe pile driven in a soft rock. Proc. 5th Int. conf. and Exhibition on Piling and Deep Foundations, Bruges, Belgium: 5.3.1–5.3.7.

Matsumoto, T., Wakisaka, T., Wang, F.W., Takeda, K. and Yabuuchi, N. (2004). Development of a rapid pile load test method using a falling mass attached with spring and damper, Proc. 7th Int. Conf. on the Appl. of Stress-Wave Theory to Piles, Selangor, Malaysia: 351–358.

Matsuzawa, K., Shimizu, S., Nakashima, Y., Isobe, Y. and Matsumoto, T. (2006). Load tests on end bearing steel H piles (Part 1: Test outline). Proc. of the Annual Meeting of Architectural Institute of Japan 2006, Yokohama: CD-ROM (in Japanese).

Middendorp, P., Bermingham, P. and Kuiper, B. (1992). Statnamic loading testing of foundation piles, Proc. of 3rd Int. Conf. of Application of Stress-Wave Theory to Piles, The Hague, Netherlands: 581–588.

Nakashima, Y., Shimizu, S., Isobe, Y., Matsuzawa, K. and Matsumoto, T. (2006). Load tests on end bearing steel H piles (Part 2: Test results). Proc. of the Annual Meeting of Architectural Institute of Japan 2006, Yokohama: CD-ROM (in Japanese).

Nishimura, S., Yamashita, K., Ogita, N., Shibata, A., Kita, N. and Ishida, M. (1998). One-dimensional stress wave simulation of rapid pile load tests, evaluation of boundary between Statnamic and dynamic loadings. Proc. 2nd Int. Statnamic Seminar, Tokyo: 337–344.

Novak, M., Nogami, T. and Aboul-Ella, F. (1978). Dynamic soil reactions for plane strain case. Jour. of Mechanical Eng. Div., ASCE, 104(EM4): 953–959.

Poulos, H.G. (1998). Pile testing – From the designer's viewpoint. Proc. of 2nd Int. Statnamic Seminar, Tokyo: 3–21.

Research Committee on Rapid Pile Load Test Methods. (1998). Research activities toward standardization of rapid pile load test methods in Japan. Proc. 2nd Int. Statnamic Seminar, Tokyo: 219–235.

Schellingerhout, A.J.G. and Revoort, E. (1996). Pseudo static pile load tester. Proc. of 5th Int. Conf. on Application of Stress-Wave Theory to Piles, Orland, 1031–1037.

Chapter 5

Influence of rate effect and pore water pressure during Rapid Load Test of piles in sand

P. Hölscher
Deltares, Delft, The Netherlands

A.F. van Tol
Delft University of Technology, Delft, The Netherlands
Department of Civil Engineering and Geosciences, Deltares, Delft, The Netherlands

N.Q. Huy
Delft University of Technology, Geo-Engineering Section, Delft, The Netherlands

SUMMARY

The research in the Netherlands focuses on the behaviour of piles which are placed with the pile toe in a sand layer. This is the mostly encountered case in the Netherlands.

Two knowledge questions are answered: the influence of rate effects in sand and the influence of pore water pressure during a rapid load test.

The influence of rate effects is studied by fast triaxial tests and model pile tests in the laboratory in dry and saturated sand. The influence on bearing capacity is limited to about 5%.

The influence of pore water pressure is studied numerically. Solving the full dynamic two phase behaviour of the problem showed that the stiffness and bearing capacity measured by a rapid load test might be over-estimated considerably. This depends on permeability of the sand. The numerical results will be verified experimentally by a centrifuge test, field tests and a demonstration project.

Finally, the research must offer tools practical to engineers for usage of the method. Therefore, a European network is created, in order to realise a standard for testing and interpretation, which fits in the framework of Eurocode 7. In order to validate the method, a database with field tests is created.

I INTRODUCTION

The Rapid Load Test on piles (RLT) has been developed originally by (Bermingham 1998). This method is nowadays known as Statnamic testing. Later, other methods for loading the pile rapidly are developed in the Netherlands by (Schellingerhout & Revoort 1996) and, more recently, in Japan (Matsumoto 2008). These developments are based on the principle of a weight falling on soft springs. In the USA and Japan, the test is considered as a regular test to determine the static capacity of piles, but in Europe the application of the test is less accepted.

A general review showed that the method might be applicable on the Dutch market when the following conditions are fulfilled: The uncertainty on rate effects and pore water pressure is solved and a generally accpeted interpretation method is available.

A Delft Cluster research on Rapid Load testing has been started, in which Delft University of Technology, Deltares and some industrial partners are working together for acceptance of the method. A CUR commission takes care of the guidance of this research and the practical applicability of the results. CUR is a Dutch organization for research and standardization in civil engineering.

One of the international working partners suggested to create regulations which are useful on an international level. This leads to the ambition to create European regulations.

2   GENERAL REVIEW OF THE RESEARCH

The research is divided in a theoretical part and a practical part.

The theoretical part points at studying the rate effects in sand and the influence of pore water pressure on bearing capacity. This part is worked out by Delft University of Technology together with GeoDelft. Laboratory experiments with fast loading tests on sand are carried out. Numerical analysis of the pore water pressure near the pile toe is made.

The developed model will be validated against centrifuge tests carried out by Deltares, a database with field tests created within this research project by the commission CUR H410. Finally, a demonstration project is foreseen, carried out by some partners.

The practical part will lead to a standard preferably a Eurocode for the execution of the test and an internationally accept guideline for the interpretation of the test. This part is carried out by Deltares, together with the commission CUR H410 and the international partners. The standard and guideline must integrate the knowledge and experience of researchers worldwide.

3   DESCRIPTION AND RESULTS OF THE RESEARCH

3.1   Introduction

In literature, the findings for rate effect show no general agreement (Huy *et al.* 2005). Figure 1 from (Huy *et al.* 2005) illustrates the reported rate effect on the vertical axis as a function of normalized loading rate at the horizontal axis.

3.2   Rate effects in sand

The laboratory research shows that in sand the rate effects are small. This conclusion is based on two laboratory tests:

– the fast triaxial tests
– the model pile tests

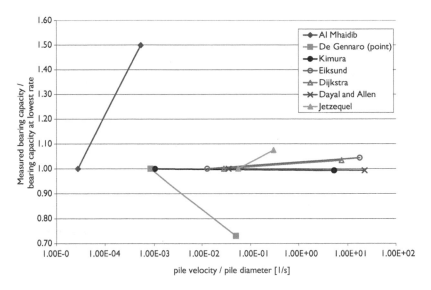

Figure 1   Overview of results for rate effect from model pile tests.

3.3   Triaxial tests

The tests have been done in a triaxial apparatus with a hydraulic loading system, which can perform both static and dynamic tests (Huy *et al.* 2005). The tests are performed on densely packed specimen with relative densities varying from 60% up to 85%. The dry specimens are prepared of oven dried Itterbeck sand in a split mold by a tamping and vibration technique. The saturated specimens are prepared on a similar way, but afterwards saturated with CO_2, which was replaced by de-aired water (Huy *et al.* 2006).

The tests are strain controlled, controlled by the downward velocity of the loading piston. In the static tests, the velocity is 0.0125 mm/s. In the dynamic tests, the velocities are 0.2 m/s and 0.55 m/s. All tests are performed until the axial strain reaches about 15%.

During the tests the applied stresses, the resulting axial strain and pore pressure are measured. The cell pressure of 100 kPa is applied by air and kept constant during all tests.

From the measured deviator stresses as a function of axial strain in static and dynamic tests with three different velocities with the same relative density (~83%), the following observations are made for dry sand:

–   The stiffness of the sand (strain =1%) shows no rate effect.
–   The peak deviator stress increases as the strain rate increases. This is due to the rate effect.

Figure 2 shows that for dry sand the angle of internal friction (on the vertical axis) increases with the density (on the horizontal axis) and with the loading rate (as a

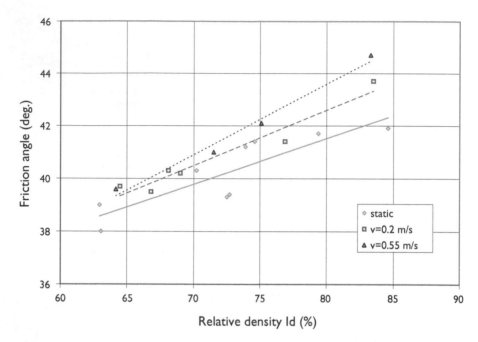

Figure 2 Friction angle on relative density for dry sand measured in triaxial tests.

parameter). The lines in Figure 2 show the trend for the static tests, the dynamic tests with $v = 0.2$ m/s and the dynamic tests with $v = 0.55$ m/s. The determined increase of the angle of internal friction with loading rate means an increase in strength of about 6%.

For saturated sand, the stiffness of the sand shows no rate effect. Figure 3 shows the maximum deviator stress with loading rate for three relative densities. As all samples cavitate during failure, the effective stress cannot be calculated. The observed strength of the sample however, shows for saturated sand no rate effect.

3.4 Model pile tests

A series of model pile tests (scale 1:10) in unsaturated sand is carried out in order to investigate the loading rate effect on the pile capacity. The tests are performed in the calibration chamber of the Geo-Engineering section of Delft University of Technology, see Figure 4. The chamber is filled with quite coarsely grained river sand. The model pile is actually a Dutch standardized CPT cone with 36 mm diameter.

The model pile is installed and statically tested by using a hydraulic rig. The model pile is dynamically tests with a drop mass of 70 kg with a drop height up to 30 cm. Between the pile head and the drop mass a series of disc springs is installed to extent the duration of the blow.

The tests showed a slight increase in tip resistance and shaft friction with increasing loading rate. Figure 5 shows the results of the tip resistance. Figure 6 shows the shaft resistance. On the horizontal axis, the pile velocity (m/s) is listed. The results of both the tip resistance and the local shaft resistance are normalized to the resistance

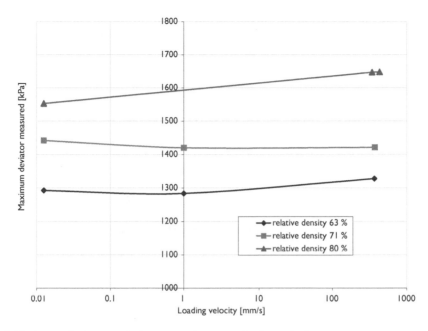

Figure 3 Maximum deviator strain in saturated tests on loading velocity.

found during the static case (1 mm/s). By measuring the soil reactions at the toe and the shaft, the inertia of the model pile is already accounted for. Therefore, only the rate effect of the soil is considered.

The increase of toe and shaft resistance is a few %, consistent with the results of the triaxial tests. This increase is however statistically not significant. Figure 5 and Figure 6 shows the mean and the 5% and 95% confidence limits, assuming the measured data is normally distributed.

The rate effect in sand is small, at maximum 5% for the loading rate used for Rapid Load tests. This conclusion is in accordance with the conclusions of most researchers, e.g. (Dayal & Allen 1975) and (Eiksund & Nordal 1996).

3.5 Pore water pressure in sand

Numerical analysis for the influence of the pore water pressure shows that these must be taken into account (Huy *et al.* 2006, Huy *et al.* 2007). First, the difference between stiffness and bearing capacity in drained and undrained sand is discussed; secondly, the partial drained case is discussed more in detail. A practical formula for the influence of the pore water pressure is presented.

The application of the RLT is still in question since the effects of excess pore pressure are not fully understood. Observations made by Hölscher (Hölscher 1995), Meada (Meada *et al.* 1998), Matsumoto (Matsumoto 1998) confirmed that during a quasi-static pile load test, the generated pore water pressure does not have enough time to dissipate sufficiently, even in sandy soil. The soil behavior in the test is not fully drained as in conventional static pile load test.

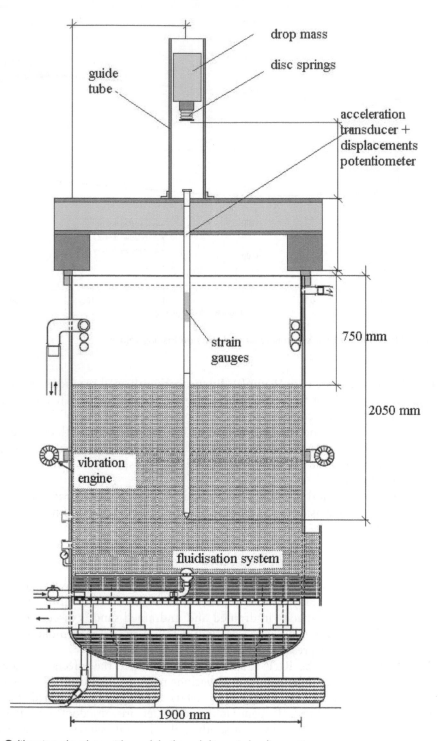

Figure 4 Calibration chamber with model pile and dynamic loading system.

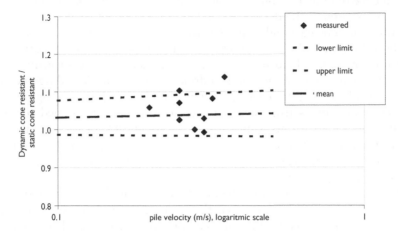

Figure 5 Tip resistance (normalised on static value) on loading rate (dashed line is 90% confidence interval).

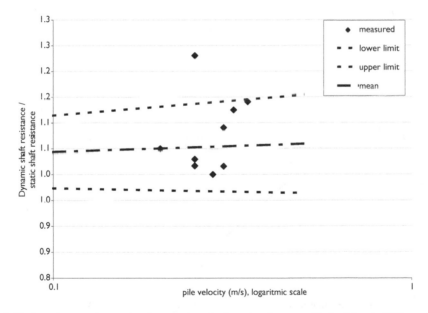

Figure 6 Shaft resistance (normalised on static value) on loading rate (dashed line is 90% confidence interval).

The RLT is simulated by dynamic finite element analysis shows that the difference between drained and undrained analysis is significant (Huy *et al.* 2006). The Plaxis model has been used for this analysis. A half sine loading of 100 ms in duration applied on pile head. The pile is embedded in sand. The soil behavior is set in two cases, fully drained and fully undrained.

The influence of drainage condition on the bearing capacity of a pile is investigated from the comparison of the derived static load-displacement curves from the simulations in different drainage conditions. The forces are calculated from the total stresses in the pile-soil interface. Figure 7 shows the static load-displacement curves from the simulation of static pile load test and the derived load-displacement diagram from the calculations in both drained and undrained condition. Any difference in these curves is induced by the difference in drainage conditions. From the figure, the static load-displacement curves from static and rapid load tests in drained condition are very similar, but the undrained simulation gives about 30% increase in static bearing capacity of the pile. Therefore, the drainage condition during rapid load testing might have effect on the static bearing capacity of a pile if the pore water pressure does not fully dissipate during the test.

In order to answer the question whether the RLT should be considered as drained, undrained or partly drained, advanced finite element simulations are made. In this analysis with the model Titan, the coupled wave propagation and consolidation are of a Biot material is solved (Hölscher 1995).

The results of the simulations (Huy *et al.* 2007) are presented as a function of the dimensionless dynamic drainage factor η. The dimensionless parameter η is defined as:

$$\eta = \frac{GT}{\gamma R^2} k \tag{1}$$

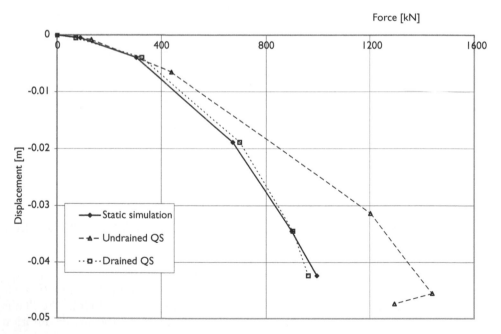

Figure 7 Difference in derived static load-displacement diagram for a fully drained and a fully undrained RLT (numerical analysis).

where G is the shear modulus (N/m^2); T is the loading duration (s); R is the pile radius (m); γ is the volumetric weight of water (N/m^3); k is the permeability (m/s). The dynamic drainage factor represents drainage condition of the test. It is related to a certain fraction of consolidation during the loading duration (Hölscher & Barends 1992).

Figures 8 and 9 show the stiffness respectively dashpot constant of a shallow foundation and the pile toe (on the vertical axis) as a function of de dynamic drainage factor η. Both stiffness and dashpot constant are normalized on the value in the fully drained case. For the drained case (high value of η), both stiffness and dashpot constant fit well with the fully drained case discussed before. The increase of the stiffness for low value of η, fits good with the increase to fully drained. The increase in dashpot constant is higher than the increase in stiffness, maybe caused by the fact that now the energy dissipation due to friction between grains and water is taken into account.

Figure 10 shows the results of simulations of an end bearing pile of 11 m in length, of which 2 m is embedded in the bearing soil layer. The soil layers are modeled with a bi-linear model; the top soil layer is very weak and the bearing layer properties correspond with dense sand.

3.6 Validation

The model, which is based on the laboratory and numerical research carried out up to now, must be validated by experiments. The most extensive part of the validation is a series of centrifuge tests. Other important steps are the validation against a database of field tests and a full-scale demonstration project. This research is at this moment on going.

A geocentrifuge offers the possibility to do scale tests on geotechnical structures. A model pile will be tested in sand at scale 1:40. The pile will be installed in flight in well-defined homogeneous sand samples by pushing into the sample. Then, the pile will be loaded statically and rapidly with different rates and force levels. The pile toe

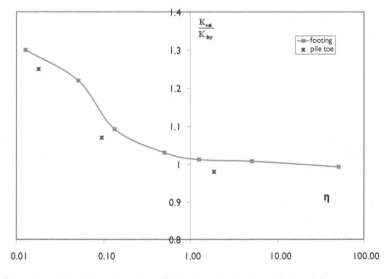

Figure 8 Stiffness (normilised) as a function of the dynamic drainage factor (η).

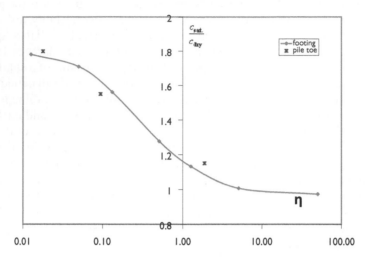

Figure 9 Dashpot constant (normilised) as a function of the dynamic drainage factor (η).

resistance, shaft friction and pore water pressure under the pile toe are measured. The test will deliver some points on the curves as shown in Figure 8 and Figure 10.

3.7   Rate factor for the unloading point method

The method Unloading point method originally suggested by Middendorp (Middendorp *et al.* 1992) seems a proper procedure to interpret the RLT. This procedure is rewritten as

$$F_{\text{static}}(t) = g_2[F(t), u(t), v(t), a(t); m, R]$$
$$F_{\text{static}}(t) = R[F(t) - ma(t) - C_4 v(t)] \tag{2}$$

with m the mass of the pile
        R a model parameter which still depends on the soil type.

The model factor R includes both the rate effects and the pore water effects. The variable C_4 has to be calculated using a prescribed algorithm (using the force in the unloading point and an averaging over the time interval between the maximum load and the maximum displacement). This can be described by

$$C_4 = \frac{1}{n} \sum_{j=1}^{n} \frac{F(t_j) - ma(t_j) - F_y}{v(t_j)} \tag{3}$$

with

$$F_y = F(t_u) - ma(t_u)$$

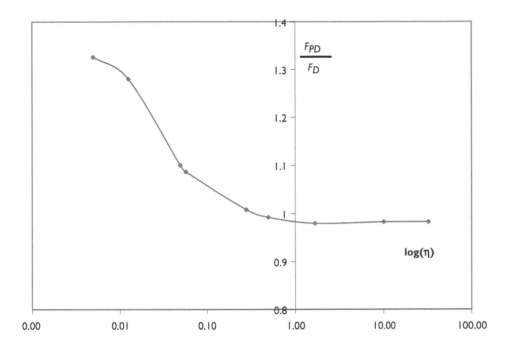

Figure 10 Ultimate toe resistance F_{PD} (normilised on the value for fully drained F_D) as a function of the dynamic drainage factor.

In these formulae t_u is the moment with $v(t_u) = 0$ (the unloading point), and the counter j counts from $j = 1$ when $F(t)$ is maximum and $j = n + 1$ occurs at t_u.

A small database with measured static and dynamic capacity is build and used to estimate the model parameter R. The database of McVay (McVay *et al.* 2003) is chosen as the starting point. It is extended with more recent cases for piles in soft soil from literature. At this moment the database contains 19 piles which did not fail during the rapid load test and 17 piles which did fail during the rapid load test. The quotient of static capacity by rapid capacity can be used as an empirical correction factor, sometimes called the rate factor. Table 1 shows the factors from the database, for both cases.

This table shows that for clay, the factor depends strongly on the occurrence of failure. For sand, a much smaller, but not significant difference between no failure and failure is observed. This result however, might depend strongly on the interpretation of the tests by the authors in literature. It is concluded that a new database, containing the measured data as a function of time, must be created.

3.8   Standard and guideline

The results of all research will be transferred to the engineering practice by creation of a standard for execution of RLT and a guideline for the interpretation of RLT. A seminar is organized to discuss the recent results of research and discuss the required content of the standard and the guideline. A reliable guideline for interpretation is only possible if the accuracy of the method is proven. This also requires a reliable

Table 1 Factors for field measurements in clay and sand.

Soil	No failure			With failure		
	No. cases	Factor	St dev.	No. cases	Factor	Soil
clay	6	0.80	0.11	6	0.50	0.07
sand	11	0.95	0.12	12	0.87	0.12

database with field measurements. This database is also needed for the derivation of partial safety factors, which should be part of the guideline. The applicability of the standard and guideline must be proven by some full-scale demonstration projects, of which one at Waddinxveen (NL) will be done within this research project.

4   CONSEQUENCES OF THE RESULTS FOR THE STANDARD AND THE GUIDELINES

The results show that:

Rate effects in sand causes that the results of a RLT overestimate the SLT results by about 6%. This is for practical usage negligible. For silt, this still have to be studied.

Pore water pressure causes that the results of a RLT might overestimate the SLT for piles in sand results up to 40%. This depends on the dynamic drainage factor, which can be included in the interpretation guideline.

The applicability of single degree of freedom models using a single rate factor must be tested more in detail by application to field data. This field data leads also to derivation of partial safety factors.

The tools described in this paper (triaxial tests and numerical analysis) can also be used to judge the consequences of rapid load tests on piles in silt.

5   CONCLUSION

The fast triaxial test is a suitable laboratory test to study the rate effects in sand. In dry sand the rate effect is observed, in saturated sand it is not clearly observed. The model pile tests in wet and saturated sand show a moderate rate effect.

The two phase dynamics modeled in finite elements is a suitable numerical tool to derive the influence of pore water pressure during RLT of piles in sand. It shows the importance of pore water build-up for piles in sand.

The existence of a large set of field measurements is essential to validate the models and to derive the partial factors for the method. Partial factors are necessary to incorporate the method in the guideline.

ACKNOWLEDGEMENT

This research is made possible by the Delft Cluster, Delft University of Technology, GeoDelft, Civil Engineering Division of the Directorate-General for Public Works and Water Management of Ministry of Waterworks and Infrastructure, IHC Hydro-hammer, Shell Global Solutions, Ballast Nedam, Volker Wessels Stevin Geotechniek, Fundex Verstraeten, IFCO Profound.

REFERENCES

Bermingham, P.D. (1998). Statnamic the first ten years, Proc. 2nd International Statnamic, Seminar, Tokyo.

Dayal, U. and Allen, J.H. (1975). The Effect of Penetration Rate on the Strength of Remolded Clay and Sand Samples. Can. Geotech. J., 12, 336, pp. 336–348.

Eiksund, G. and Nordal, S. (1996). Dynamic model pile testing with pore pressure measurements. Proc. 5th Int. Conf. Appl. Stress-Wave Theory to Piles, Orlando, Sept. 1996, Gainesville, Univ. Florida, Dep. Civ. Eng., pp. 1–11.

Hölscher, P. (1995). Dynamical response of saturated and dry soils, Ph.D. thesis, Delft University of Technology.

Hölscher, P. and Barends, F.B.J. (1992). The relation between soil-parameters and one-dimmensional toe-models, Proc. 4th Int. Conf. Application of Stress Wave Theory to Piles, the Netherlands, pp. 413–419.

Huy, N.Q., Dijkstra, J., van Tol, A.F., and Hölscher, P. (2005). Influence of loading rate on the bearing capacity of piles in sand, in proc. 16th Int. Conf. Soil Mech. Geotech. Eng., Osaka, Sept. 2005, Rotterdam, Millpress, Vol.4, pp. 2125–2128.

Huy, N.Q., van Tol, A.F. and Holscher, P. (2006). Laboratory investigation of the loading rate effects in sand, report of TU-Delft, 24-08-2006.

Huy, N.Q., van Tol, A.F. and Holscher, P. (2007). A numerical study to the effects of excess pore water pressure in a rapid pile load test, draft paper for ECSMGE 14th, Madrid, Spain, September 2007.

Huy, N.Q., van Tol, A.F. and Holscher, P. (2006). Numerical simulation of quasi-static pile load test, Proc. 10th Int. Conf. Piling Deep Foundations, Amsterdam, June 2006, London, Emap, pp. 677–683.

Matsumoto, T. (1998). A FEM Analysis of a Statnamic Test on Open-ended Steel Pipe Pile. Proc. 2nd. Int. Statnamic Seminar, Tokyo, pp. 287–294.

Matsumoto, T. (2008). Practice of rapid load testing in Japan, Rapid Load Testing in piles, Hölscher P., Van Tol, A.F. (eds) Francis Taylor, September 2008.

McVay, M., Kuo, C.L. and Guisinger, A.L.(2003). Calibrating resistance factors for load and resistance factor design for Statnamic load testing, report University of Florida, 491-045-048-2312, March 2003.

Meada, Y., Muroi, T., Nakazono, N., Takeuchi, H. and Yamamoto, Y. (1998). Applicability of Unloading-point-method and signal matching analysis on the Statnamic test for cast-in-place pile. Proceedings of the 2nd International Statnamic Seminar, pp. 99–108.

Middendorp, P., Bermingham, P. and Kuiper, B. (1992). Statnamic load testing of foundation piles, In: "Proc. 4th Int. Conf. Appl. Stress-Wave Theory to Piles, The Hague, Sept. 1992", Rotterdam, Balkema, pp. 581–588.

Schellingerhout, A.J. and Revoort, E. (1996). Pseudo static pile load tester, Proc. 5th Int. Conf. Appl. Stress-Wave Theory to Piles, Orlando, Sept. 1996, Gainesville, Univ. Florida, Dep. Civ. Eng.

Chapter 6

Concrete stress determination in Rapid Load Tests

Michael Stokes and Gray Mullins
University of South Florida, Tampa, FL, USA

SUMMARY

This paper reviews the effects of loading rate and stress level on concrete strength and elastic modulus as it pertains to the regression of rapid load test data. Based on detailed laboratory research, a stress-strain-strain rate relation is developed then normalized for application to various concrete mix designs. The model is validated for two field tests. For long piles subjected to significant amounts of side shear, it is concluded that the non-linear behavior of concrete must be taken into account for a proper interpretation of load distribution. A practical method is presented herein.

I INTRODUCTION

The effects of loading rate on concrete strength and elastic modulus have long been recognized with regards to seismic and/or dynamic loading events. In such cases an increase in strength is reported to show up to 15 percent higher values. Rapid load tests of deep concrete foundation elements use the concrete modulus determined from field calibration via strain gages directly under the applied (known) load. Therein, rate-dependent effects can be directly incorporated into the determination of load from strain measurements elsewhere in the foundation element. However, in many cases the strain rate at various strain gage levels throughout the foundation may differ and the magnitude of strain may or may not lend itself to linearly approximated stress-strain relationships.

Hence, the modulus selected to determine the load sensed at these levels can produce erroneous results if a single value is assumed. This article presents the findings of an experimental program designed to quantify the relationship between concrete stress, strain, and strain rate. Developed equations are used to show the effect of various stress evaluation methods on rapid load test results.

2 GENERAL REVIEW OF THE RESEARCH

In long test piles that exhibit a significant amount of side shear, multiple strain gages are often embedded at different levels so as to better understand the distribution of forces along the length of the pile (Figure 1). Strains recorded at these gage locations lend insight into the amount of load transferred to the surrounding soil. In order to perform these computations, the cross-sectional area and elastic modulus at each gage location are typically used to estimate the load at that level. In the case of the

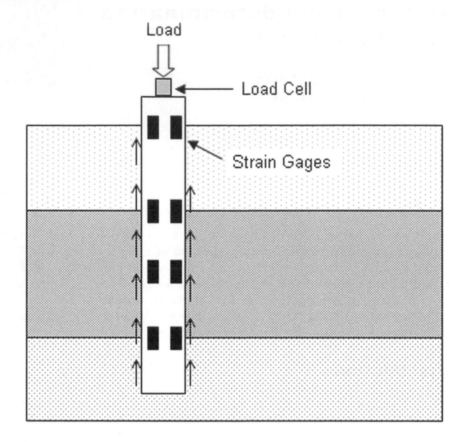

Figure I Instrumentation scheme in a long pile.

cross-sectional area, it can be accurately measured prior to driving (driven piles) or legitimately assumed based on the borehole diameter (drilled shaft). The modulus at each gage location is taken as a composite modulus based on the modular ratio and the proportions of steel and concrete. Where the elastic modulus of steel is well documented and remains relatively constant regardless of its grade (200 GPa or 29000 ksi), information gathered from a concrete compression test (ASTM C39) on a sample of the mix and empirical equations are often used to estimate the elastic modulus of concrete. The following equations from the American Concrete Institute (ACI) are based on a secant modulus of approximately 0.45 f_c' (Eqns. 1 and 2) (for symbols, see page 100).

$$E_c = 33w_c^{1.5}\sqrt{f_c'} \quad (w_c \text{ in pcf and } f_c' \text{ in psi}) \tag{1}$$

$$E_c = 57000\sqrt{f_c'} \text{ (for normal weight concrete)} \tag{2}$$

Alternately, the composite modulus can be calculated directly by calibrating the measured load at the top of the pile with near-surface strain or modulus gages.

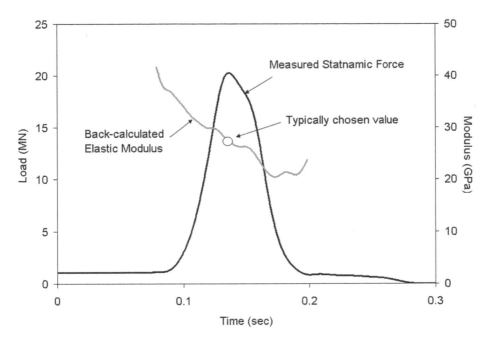

Figure 2 Back-calculated (field calibrated) elastic modulus from near-surface or modulus gages.

This process of "matching" the top level strain to the load eliminates the need to determine the individual contribution of the steel and concrete, for the calibration inherently accounts for the composite cross-section. Adjustments are then made depending on the reinforcement ratio throughout the length of the pile. If the measured load and strain are used to back-calculate a composite modulus as a function of time, it becomes apparent that the modulus changes significantly throughout the test (Figure 2).

Whether using empirical equations or near-surface gages, the composite modulus is assumed to remain constant throughout the duration of the load test, and a simple linear relationship (Hooke's Law) is used to relate the measured strain to the stress at a particular gage level (Figure 3). This relationship may be valid as it applies to the steel, but the nonlinear behavior of concrete can lead to grossly inaccurate estimations of stress within the concrete portion of the composite section. A more representative estimation of stress can be computed with the implementation of a nonlinear model where the concrete modulus varies as a function of the level of strain.

Recently, consideration has been given to the use of a variable strain-dependent modulus in the regression of static load test data (Fellenius 2001). In this particular case study, static load test data on a 20 m monotube pile was evaluated using both a strain-dependent modulus and a constant, average modulus. Results from the evaluation indicated that the mid-level pile stresses were lower than what was computed using a constant modulus, and lower level stresses were higher. It was concluded that if the constant modulus were used, the "resistance acting between the two levels would have been determined with an about 10 percent to 20 percent error."

The Hognestad model is one of the more popular parabolic stress-strain relationships whose first derivative, or modulus, varies as a function of strain. It expresses a

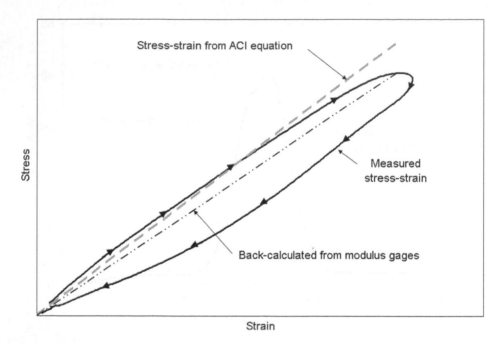

Figure 3 Assumed linear stress-strain relationship *vs.* measured response.

stress-strain curve that is only dependent on the compressive strength and the corresponding ultimate strain (Eqns. 3 and 4). Though the use of a nonlinear stress-strain model may account for a strain-dependent modulus, it does not address unloading stresses or compensate for load rate effects that occur during rapid and/or dynamic load tests.

$$f_c = f_c' \left[\frac{2\varepsilon_c}{\varepsilon_o} - \left(\frac{\varepsilon_c}{\varepsilon_o} \right)^2 \right] \quad \text{or} \tag{3}$$

$$f_c = -\frac{f_c'}{\varepsilon_o^2} \varepsilon_o^2 + \frac{2f_c'}{\varepsilon_o} \varepsilon_c \tag{4}$$

The effect of increased load rates on the compressive strength of concrete has long been realized. Takeda and Tachikawa (1971) proposed a relationship between stress, strain, and strain rate from test results of 14 different batch mixes with varying aggregate sizes subjected to strain rates ranging from 1 me/s to 1 e/s (Figure 4). Despite the different mix designs and concrete properties, all specimens produced geometric similarities in their stress-strain-strain rate relationships. Most researchers agree that the compressive strength of concrete increases as much as 15% at strain rates of 0.02 e/s, although the increase in modulus is more moderate (Fu *et al.* 1991). Some signal matching algorithms for dynamic wave analyses recognize these rate effects and compensate by varying the concrete modulus throughout the length of the pile.

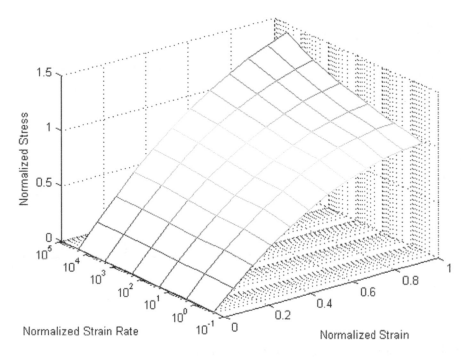

Figure 4 Normalized concrete stress-strain-strain rate relationship (adapted from Takeda and Tachikawa (1971)).

However, the modulus assigned to each level is again assumed constant and Hooke's Law used to determine the stress.

Strain values in a pile exhibiting either elastic behavior or large amounts of side shear can vary significantly throughout the length of the foundation (location) as well as throughout the loading event (time). During rapid and dynamic tests, these strain measurements are almost never in phase and therefore add a degree of difficulty when determining the ultimate capacity. Consider the strain at multiple levels throughout the drilled shaft in Figure 5. Not only is it evident that the upper level gages experience higher strains, but they undergo much larger strain rates than the lower level gages. Also prominent is a delay of maximum strain, or phase shift, at each level. Had the maximum strains occurred simultaneously at each gage level, then perhaps compensation for strain rate effects may not prove to be worthwhile when calculating the ultimate capacity, since the strain rate at maximum strain is zero. However, the ultimate capacity in rapid and dynamic tests usually occurs at a point in time between the maximum strains of the upper and lower levels. Because of this, strain rate effects must be considered, and unloading stresses must be accurate.

3   CONSEQUENCES OF THE RESULTS FOR THE STANDARD AND THE GUIDELINES

The nonlinear hysteretic behavior of concrete coupled with its sensitivity to increased loading rates suggests that the use of a constant concrete modulus and a linear

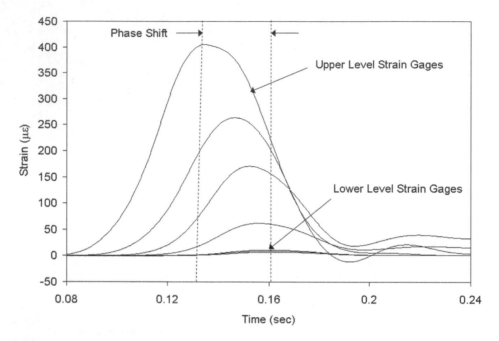

Figure 5 Strain distribution throughout a pile during a rapid load test.

stress-strain relationship may be inappropriate for certain load tests that exhibit a wide range of strains and strain rates throughout a pile. Previous research on static load test results has shown that the use of a constant concrete modulus in determining gage level stresses from measured strains can substantially over-predict stresses near the top and under-predict stresses near the bottom of a pile. With this in mind, strain rate effects introduced during rapid load tests only compound the complexity. Regardless of the type of load test used or analysis procedure performed, provisions for a variable modulus and/or nonlinear hysteretic stress-strain-strain rate model may prove to refine the test results and ultimately yield a more accurate interpretation of the foundation performance. A suitable solution is provided herein.

4   DESCRIPTION AND RESULTS OF THE RESEARCH

In order to confirm the results of previous researchers and identify the strain rate effects indicative of rapid load tests, thirty 50 mm × 100 mm (2 in × 4 in) cylindrical mortar specimens were cast and tested at various loading rates using a MTS 809 Axial/Torsional Test System and a 180 kN (20 T) laboratory-scale rapid load testing device (Stokes 2004). Due to the limitations of the rapid load testing device, smaller diameter mortar cylinders were chosen as test specimens to ensure that failure was achievable, especially with an anticipated rate-dependent strength increase of 15%. Though aggregate scaling issues existed between the mortar specimens and published concrete data, results of Takeda and Tachikawa indicate that their behaviors should be similar.

Table 1 Specimen mix design.

Batch volume	0.014 m³
w/c ratio	0.485
Mini-slump pat Diameter	63.61 cm²

Table 2 Specimen load rates.

Number of specimens	Load rate (kN/sec)
4	0.7
2	2.4
2	7.7
3	25.4
3	82.3
3	266.3
3	861.6
3	2788.1

Table 1 contains the mix design for the batch of specimens. Each specimen was instrumented with 2–10 mm resistive type strain gages located at mid-height and 180 degrees apart. Data was collected using a data acquisition device at various sampling rates appropriate for the loading rate of each specimen.

Phase 1 testing   The first 23 cylinders were tested to failure using the MTS compression device at load rates that ranged from 0.7 to 2788 kN/sec (Table 2). A minimum of two cylinders were tested at a given load rate as is customary with ASTM standards. Scaled testing caps were made to replicate those used in larger cylinder tests and reduce the frictional restraining forces between the ends of the specimens and the testing platens.

Phase 2 testing   The remaining 7 cylinders were tested using the laboratory-scale rapid load testing device. This device was used to achieve representative rapid load test strain rates that were beyond the limitations of the MTS device. Though the intent was to load each cylinder to failure, 3 of the tests did not provide adequate force to break the cylinders, therefore unload data was inadvertently obtained. This mishap proved fortunate in the later developmental stages of the stress-strain model.

4.1   Model formulation

Results from the first testing phase and the cylinder breaks from the second testing phase showed that the stress and modulus at any given strain increased with increasing load rate (Figure 6). Though the strength exhibited a distinct increase, any trend in the ultimate strain was indiscernible. If assumed that the strain rate remains constant throughout the test as is loading rate, then the individual tests can be plotted in a

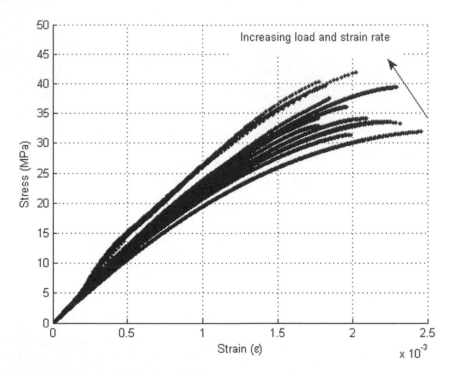

Figure 6 Stress-strain relationship from first and second phase laboratory tests.

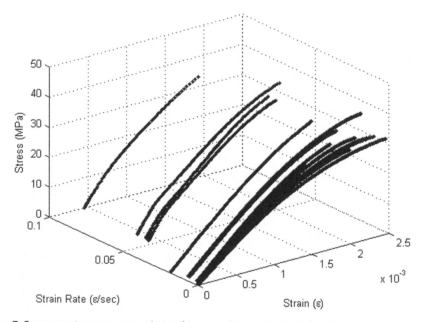

Figure 7 Stress-strain-strain rate relationship assuming constant strain rates.

3-dimensional space as seen in Figure 7. In order to determine the best-fit surface through these data points, the Hognestad parabolic stress-strain model was used to fit the two tests run at the upper ASTM loading rate limit (0.7 kN/sec). This well-accepted parabolic relationship proved to be a relatively simple but reliable model for fitting the ASTM load rate data (Figure 8).

To better understand the strain rate effects and to expand the Hognestad relationship into the strain rate dimension, all data falling within constant values of strain (strain bands) were plotted in the stress-strain rate plane (Figure 9). From this, a cubic relationship for each constant-strain band was determined and applied to the base Hognestad model such that a best-fit surface could be created through the data (Figure 10). A more familiar surface appears when the strain rate is converted to log-scale as is traditionally published (Figure 11).

This model was developed with the assumption that both the strain rate and load rate remained constant throughout each test. However, examination of the data as a function of time revealed that they varied significantly. Variations in the loading rate are largely due to the logic statements in the computer code and feedback controls attempting to hone the testing device on a specified loading rate. Although variations in the strain rate are partly a result of the load rate control, they are mostly the result of the nonlinear behavior of the mortar. As previously discussed, a simple linear stress-strain model does not adequately describe the nonlinear stress-strain response of concrete and/or mortar, therefore it should not be expected that the time derivatives be directly proportional.

$$f_c \not\propto \varepsilon_c \quad \therefore \dot{f}_c \not\propto \dot{\varepsilon}_c \tag{5}$$

If an instantaneous strain rate is calculated for each test instead of assuming that the strain rate remains constant, a slightly different plot emerges (Figure 12). After performing the same procedure described above for determining the best-fit surface, a model was developed which exhibits a similar shape but more subtle increase in ultimate strength with increasing strain rate (Figure 13).

$$f_c = -\frac{f_c'}{\varepsilon_o^2}\varepsilon_c^2 + \frac{f_c'}{\varepsilon_o}\varepsilon_c \left[\frac{5}{4}(\dot{\varepsilon}_c - \dot{\varepsilon}_o)^{\frac{1}{3}} + 2\right] \tag{6}$$

Upon close examination, it can be seen that the above equation is a modified version of the published Hognestad relationship. The modification is in the form of a strain rate multiplier applied only to the "B" coefficient of the base parabolic equation. As a result of the data analysis, it was determined that modifications to the "A" coefficient led to no sizeable effects that could not be accounted for entirely in the "B" coefficient. A statistical analysis was performed against the raw data and their corresponding modeled values which yielded a coefficient of determination of 0.993. Again, the modeled surface displays a familiar geometric shape when the strain rate axis is plotted in log-scale (Figure 14).

Another attempt at developing a modeled surface was made; however, this model was based on a logarithmically increasing stress as a function of strain rate (Eqn. 7).

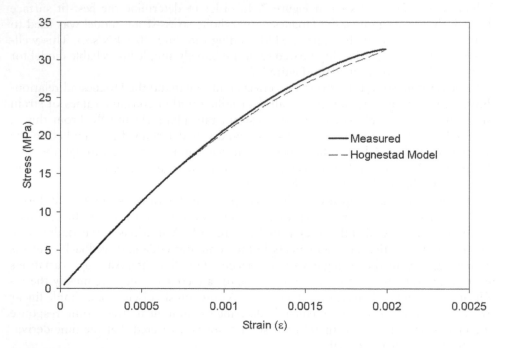

Figure 8 Hognestad modeled stress-strain relationship for ASTM strain rate data.

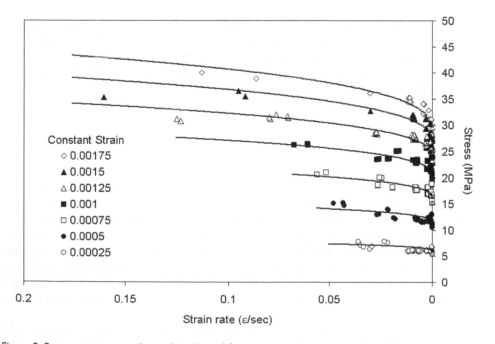

Figure 9 Stress-strain rate relationship plotted for incremental constant values of strain.

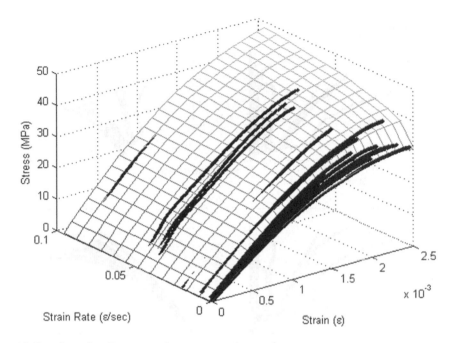

Figure 10 Best-fit surface for assumed constant strain rate data.

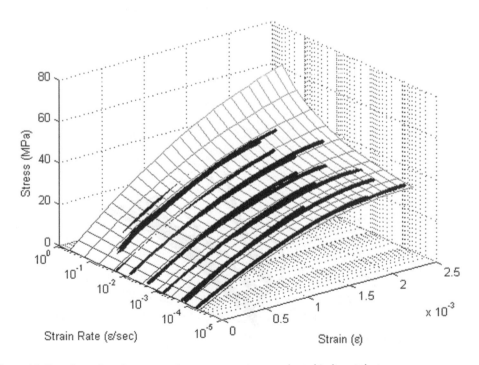

Figure 11 Best-fit surface for assumed constant strain rate plotted in log scale.

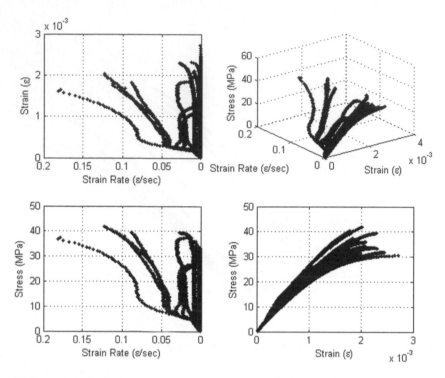

Figure 12 Data plotted using instantaneous strain rate instead of assumed constant strain rate.

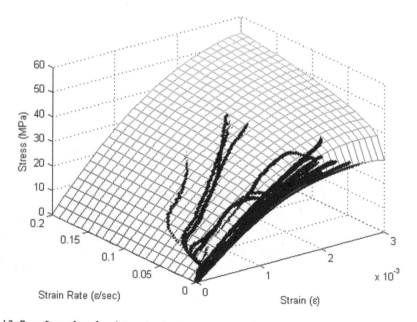

Figure 13 Best-fit surface for data using instantaneous strain rate.

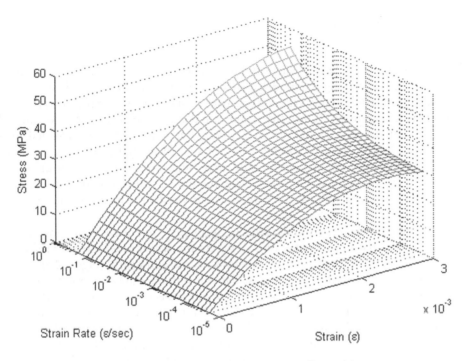

Figure 14 Best-fit surface for data using instantaneous strain rate (log scale).

Though the logarithmic-based multiplier promised a better fit at lower strain rates, it did not represent the increase in stress beyond 0.2 e/sec well. Despite this downfall, the logarithmic model yielded a coefficient of determination of 0.992, and although both models seemed to match well, the cubic model was chosen due to its consistency with published results.

$$f_c = -\frac{f_c'}{\varepsilon_o^2}\varepsilon_c^2 + \frac{f_c'}{\varepsilon_o}\varepsilon_c\left[\frac{1}{17}\ln\left(\frac{\dot{\varepsilon}_c}{\dot{\varepsilon}_o}\right)+2\right] \tag{7}$$

In an attempt to apply the model toward all concrete/mortar types, the base Hognestad model and the strain-rate compensating multiplier were normalized to all three input parameters: the ultimate strength (f_c'), the strain at ultimate strength (ε_o), and the strain rate at which the ultimate strength and strain were determined $(\dot{\varepsilon}_o)$. All values can be determined from a standard concrete cylinder compression test (e.g. ASTM C39). Though it has been noted that the strain rate and load rate do not remain constant throughout a compression test, the ASTM standard makes provisions for this variability by stating that "the designated rate of movement shall be maintained at least during the latter half of the anticipated loading phase of the testing cycle." For the purposes of normalization, the strain rate is taken near the latter half of the testing cycle where it remains fairly constant.

$$f_N = 2\varepsilon_N \left[\frac{1}{50} (\dot{\varepsilon}_N - 1)^{\frac{1}{3}} + 1 \right] - \varepsilon_N^{\,2} \tag{8}$$

When plotting a single test where failure occurred, the stress path can be seen as it maneuvers closely along the modeled surface (Figure 15). However, in a test specimen where failure did not occur, the stress path deviates from the surface as the strain rate decreases (Figure 16). Once the strain rate decreases to within ASTM rates, the stress does not decrease as the model predicts, but remains higher. This led to the presumption that the model was only valid for specimens that were loaded to failure where the strain rate was in a continually increasing state. For tests that did not experience failure but underwent a load cycle, the model sufficiently predicts the stress up to the point of maximum strain rate; but afterwards, the measured stress follows a different path.

By plotting the test data from the Phase 2 specimens that underwent a load cycle, it was determined that beyond the point of maximum strain rate (hereafter referred to as the transition point), constant strain bands decreased linearly (Figure 17).

A second surface was fit to model these data beyond the transition point (Figure 18). It is important to note that the origin for the linear decrease of the second model is entirely based on the stress-strain relationship defined at the location of the transition point. Depending on the magnitude of loading, the stress-strain relationship diverges from a common loading surface beyond the transition point (Figure 19). Prior to reaching the transition point, the A and B coefficients of the parabolic stress-strain curve are changing based on the cubic model (equation 8). When the transition point is reached, the strain rate multiplier on the B coefficient remains constant for the remaining portion of the test, and a linearly decreasing offset is applied to the entire relationship (equation 9). This offset is a function of strain-rate and was derived using only data in the decreasing strain-rate portion of the test. Despite a second inflection in strain rate near the end of the loading cycle (Figure 20), this linearly offset model sufficiently predicts the end-of-test stresses.

$$f_N = 2\varepsilon_N \left[\frac{1}{50} (\dot{\varepsilon}_{N-tp} - 1)^{\frac{1}{3}} + 1 \right] - \varepsilon_N^2 - 0.00012 \left(\dot{\varepsilon}_{N-tp} - \dot{\varepsilon}_N \right) \tag{9}$$

Figures 21 and 22 offer a three-dimensional outlook on the normalized stress path taken by one of the specimens (STN 6) and the modeled stress as determined by the pre-transitional cubic model and the post-transitional linear model (Figure 23). The coefficient of determination is 0.997, thereby showing that there is good agreement between the measured and modeled stress.

4.2 Case studies

During the development stage of the model, input values were normalized with respect to the ultimate compressive strength, the strain at ultimate strength, and the strain rate at which these values were determined. The intent was to define a model that could be applied to the regression of concrete pile load test data despite the many possible

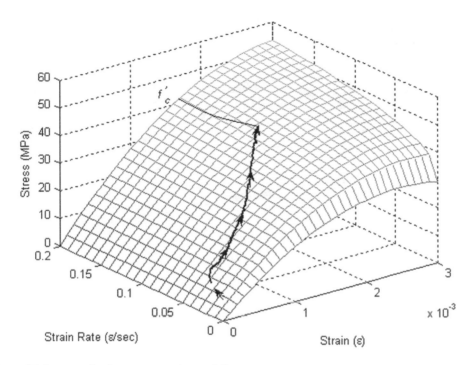

Figure 15 Stress path of a specimen taken to failure.

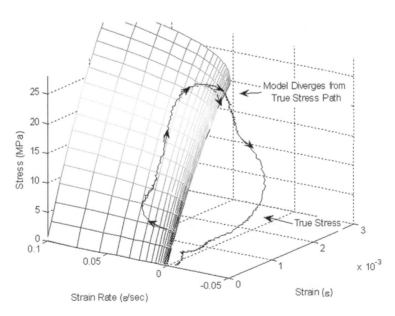

Figure 16 Stress path of a specimen that underwent a load cycle and divergence from original surface.

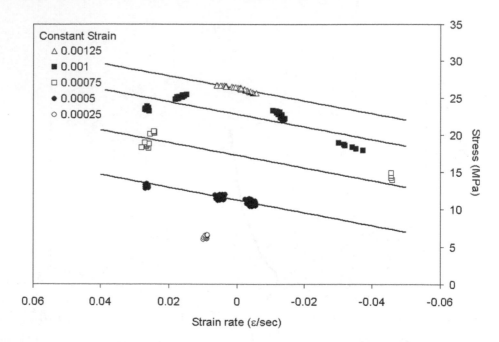

Figure 17 Constant strain bands for load cycle specimens.

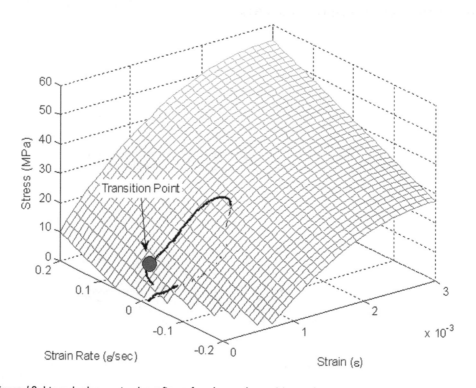

Figure 18 Linearly decreasing best-fit surface beyond transition point.

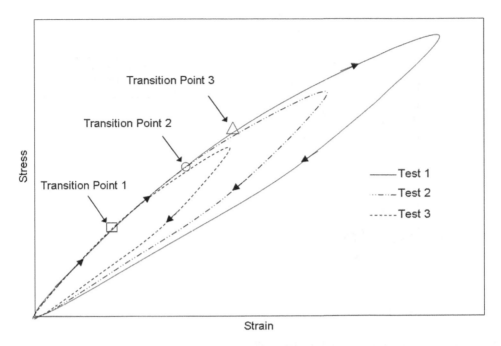

Figure 19 Dependence of unloading response on degree of loading.

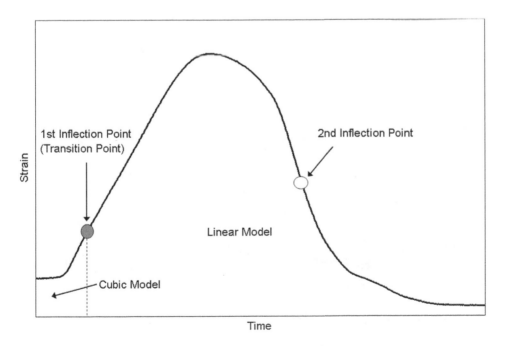

Figure 20 Regions of model application and locations of strain inflections.

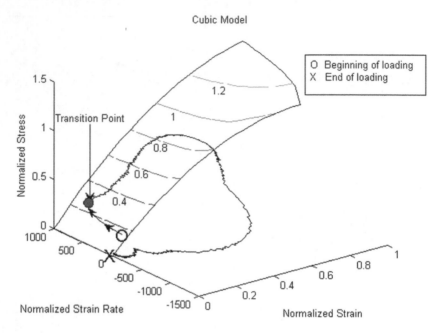

Figure 21 Normalized pre-transitional cubic model.

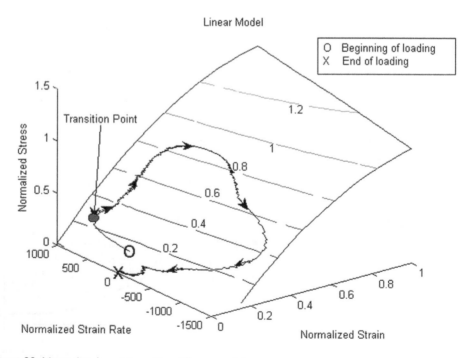

Figure 22 Normalized post-transitional linear model.

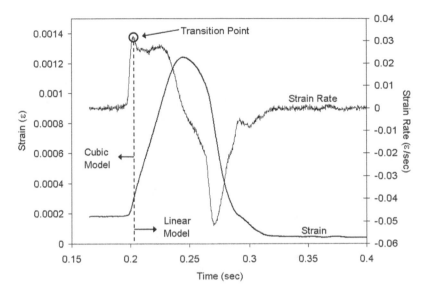

Figure 23 Transition point defining the application of the cubic and linear models.

variations in the concrete mix design. In order to test the viability of the model, two case studies were examined where stress, strain, and strain rate information were available. These case studies were selected due to the presence of modulus gages and concrete strength information. In both cases, the upper level strain gages were located such that there was no appreciable loss of load due to load shed. This ensured that the true axial stress at the gage location could be reasonably computed and provide a means for comparison.

Houston Shaft S-1   As part of a collaborative effort between the University of South Florida (USF) and the University of Houston (UH) to demonstrate the effectiveness of post grouting, four drilled shafts were constructed then tested using a 16 MN statnamic load testing device (Mullins and O'Neill 2003). Two of the 1.2 m (4 ft) diameter drilled shafts were embedded in sand, and the other two were embedded in clay. One shaft from each set served as a control (ungrouted), while the other was post grouted prior to load testing. For the purposes of this article, the particular shaft of interest is the ungrouted control embedded in sand (Shaft S-1).

Shaft S-1 was constructed to 6.4 m (21 ft) using 18-#9 reinforcing bars with #4 ties on 15 cm (6 in) centers. Strain gages were placed on the reinforcing cage so as to correspond to changes in the soil strata at 1.2 m (4 ft), 2.1 m (7 ft), and 6.1 m (20 ft). Each strain gage level consisted of four resistive-type electrical strain gages positioned 90 degrees apart.

Since the upper level strain gages were located close enough to the ground surface to reasonably assume no losses due to load shed, the measured statnamic force was used to determine the concrete stress at the upper level gage location.

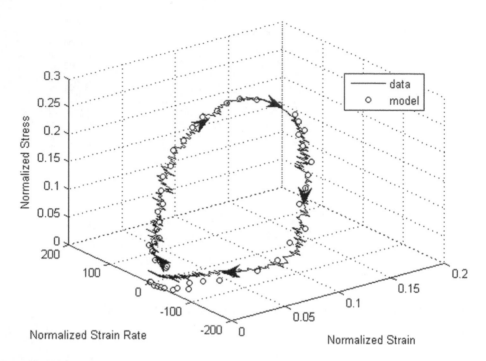

Figure 24 Modeled stress path *vs.* true measured stress path for Shaft S-1.

Because the force was being applied to a composite cross-section, a portion of the load had to be discounted due to the presence of the reinforcing steel. This was accomplished by using the measured strain, known cross-sectional area of steel, and Hooke's Law to determine the force in the steel. Also, inertial effects were considered since the load cell and the strain gages were separated by a 3518 kg mass of concrete and steel. After these adjustments were made, the corrected force was divided by the cross-sectional area of concrete to obtain the stress at the upper gage level.

Using the information from the concrete cylinder compression tests, the ultimate strength (26.2 MPa or 3.8 ksi), ultimate strain (0.001862 e), and strain rate (40 me/sec) were determined and the data normalized. Figure 24 shows the normalized data plotted with the modeled response.

Good agreement was found between the data and the model despite the differences in the mix design between the concrete drilled shaft from which the data was taken and the mortar cylinders from which the model was developed (coefficient of determination of 0.980).

Bayou Chico Pier 15   The Bayou Chico bridge project involved the replacement of an existing drawbridge in Pensacola, Florida with a newer high-rise bridge. One of the 600 mm square prestressed piles located at Pier 15 of the new bridge became part of an extensive load test program implemented by the Florida Department of Transportation (FDOT) that compared cycles of static, statnamic, and dynamic load tests (Lewis 1999). Prior to casting, vibrating wire and resistive type strain gages were mounted 180 degrees apart at five levels throughout the pile: 0.71, 2.57, 4.55, 6.6, and 8.53 m from the toe. The pile was driven to a depth of 8.4 m which resulted in the uppermost gages being positioned above ground surface.

A dynamic test was recorded on the pile during the last blow of installation, after which three consecutive static load cycles were performed. Nearly two months later, a statnamic test was completed using a 14 MN device. For the purposes of this paper, only results from the statnamic test are considered.

An inertial correction was applied to determine the stress at the upper level strain gages and similar calculations performed as outlined in the previous case study; however, no correction was made on behalf of the prestressing strands. Since the strands remain in tension throughout the load test, any compressive stresses imparted by the device are taken entirely by the cross-sectional area of concrete. The only compensation for prestressing was applied as an initial compressive offset at strand release of 7.6 MPa.

No concrete cylinder compression test information was readily available for this particular case study, therefore information from previous data regressions was used to determine the ultimate strength (49.3 MPa or 7.15 ksi). The ultimate strain (0.0027ε) and strain rate (40 $\mu\varepsilon$/sec) were assumed based on cylinder compression test results of similar piles. Figure 25 shows the normalized data plotted with the modeled response. Again, good agreement is notable between the data and the model (coefficient of determination of 0.979).

When plotting the normalized results of both case studies, the difference in pile type is accentuated by the prestressing offset visible in the Bayou Chico pile (Figure 26).

4.3   Model application

Since the results from the case studies indicate that the model provides a reasonably accurate prediction of upper level gage stresses in both the cast-insitu drilled shaft and the precast driven pile, the model was applied to a load test scenario to determine the effect on the regressed data. The scenario involves a relatively long drilled shaft (length to diameter ratio of 49) tipped in soft soil but which exhibits a significant amount of side shear when rapidly loaded to geotechnical failure. These conditions highlight the need to address variations in strain magnitude and rate throughout a given test shaft.

The hypothetical shaft is 1.22 m (4 ft) in diameter and 61 m (200 ft) in length with strain gages located at ground surface and depths of 15 m (49 ft), 30 m (98 ft), 55 m (180 ft), and 60 m (197 ft). Longitudinal reinforcement is continuous throughout the length of the shaft such that the cross-sectional area of steel and concrete at all gage levels is 117 cm^2 (18 in^2) and 1.156 m^2 (12.44 ft^2) respectively. The concrete strength (f_c'), ultimate strain (ε_o), and strain rate at which the compression test was performed ($\dot{\varepsilon}_o$) is 27.58 MPa, 0.0027 ε, and 40 $\mu\varepsilon$/sec.

The measured load test data is presented in Figure 27. Variations between the strains at each gage level implies that a large portion of the applied load is taken by side shear. Also predominant is the characteristic phase lag between peak strain values seen in rapidly loaded piles. Both phenomena lend the data to a segmental analysis (Mullins *et al.* 2002). However, to perform the analysis, it is necessary to determine the force or stress applied at each gage level.

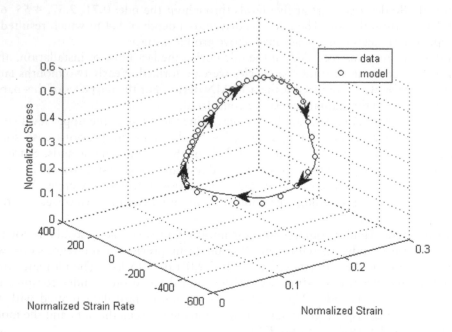

Figure 25 Modeled stress path *vs.* true measured stress path for Bayou Chico Pier 15.

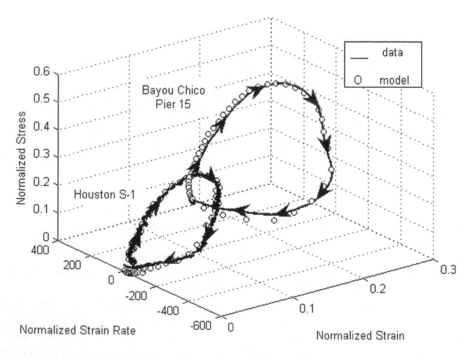

Figure 26 Individual Houston S-1 and Bayou Chico stress paths showing the stress offset due to prestressing.

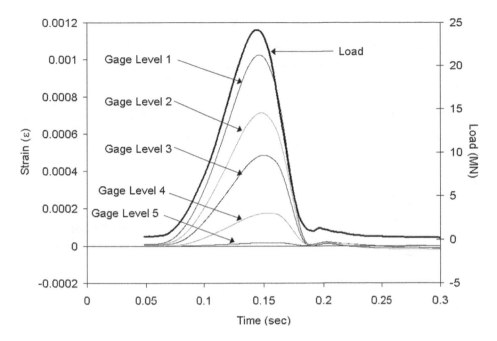

Figure 27 Hypothetical rapid load test data.

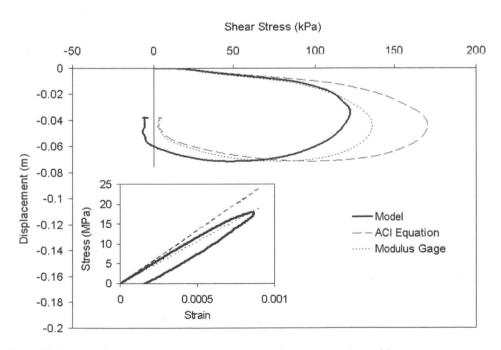

Figure 28 Segment I dynamic t-z curves and accompanying stress-strain model.

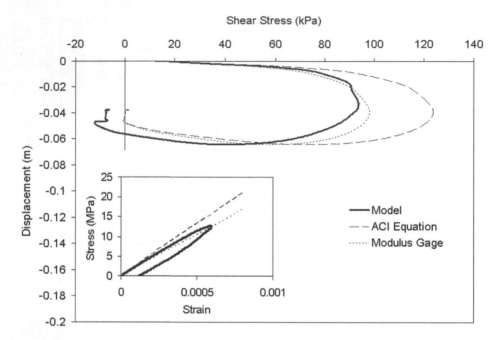

Figure 29 Segment 2 dynamic t-z curves and accompanying stress-strain model.

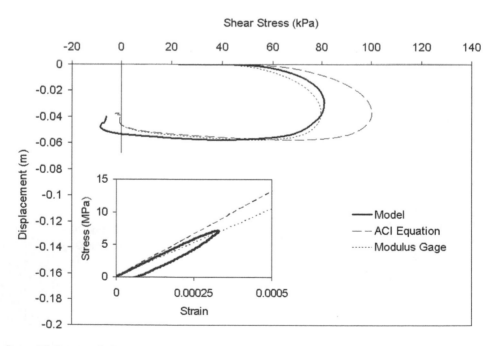

Figure 30 Segment 3 dynamic t-z curves and accompanying stress-strain model.

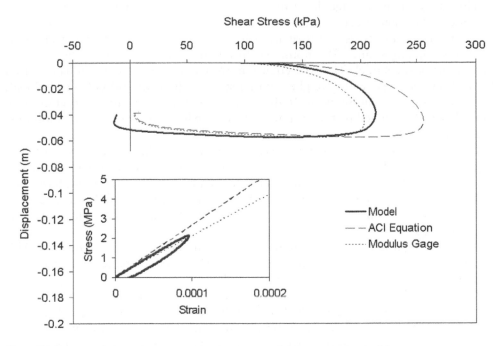

Figure 31 Segment 4 dynamic t-z curves and accompanying stress-strain model.

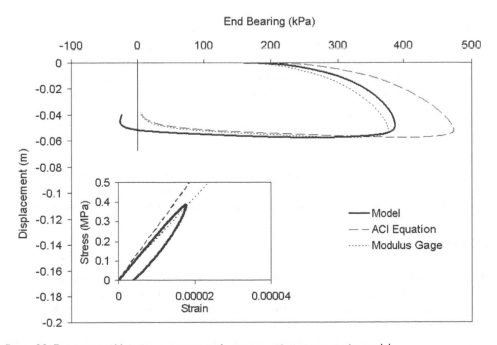

Figure 32 Dynamic end bearing response and accompanying stress-strain model.

The stress at each gage location within the shaft is computed using three methods: (1) a linear elastic stress-strain relationship based on an ACI modulus, (2) a linear elastic stress-strain relationship based on a back-calculated modulus from the top gages, and (3) a nonlinear hysteretic stress-strain relationship based on the proposed model. The ACI equation produces a concrete modulus of 24.68 GPa (3580 ksi), that when weighed with a steel modulus of 200 GPa (29000 ksi), yields a composite modulus of 26.44 GPa (3834 ksi). The calibrated composite modulus as determined by the ground surface gage level (back-calculated modulus) is 21 GPa (3046 ksi).

Once the stress/load at each gage level is computed, the difference in load between gage levels is divided by its corresponding segment surface area to obtain dynamic t-z curves which are uncorrected for inertia and damping effects (Figures 28 through 32). Segment 1, the portion of the shaft between gage level 1 and 2, shows a 21% reduction in maximum stress when the back-calculated modulus is used instead of the ACI modulus.

All other segments, including the toe of the shaft, show the same reduction. This constant reduction in load is a direct result of a 21% reduction between the back-calculated and the ACI calculated composite moduli.

When considering the segment loading based on the modeled stress, a distinct shift in the distribution of load along the length of the shaft becomes noticeable. The upper level segments (1 and 2) do not distribute as much load as the back-calculated modulus curve suggests, whereas the lower level segments (3 and 4) distribute more.

This occurrence can be explained using the segment stress-strain plots located within each figure (Figures 28–32).

When observing the stress-strain relationship of segment 1 (Figure 28), although barely distinguishable, the linear modulus gage relationship intersects the modeled relationship slightly above the point of maximum stress. As the magnitude of strain and strain rate decreases (moving downward through the shaft), the point of intersection begins to fall below the point of maximum stress on the model (Figures 29–32). This is again a by-product of the nonlinearity of the concrete material and results in the overestimation of stresses in the upper segments and underestimation of stresses in the lower segments.

At this point, a segmental unloading point analysis (SUP) could be performed on the data to obtain the equivalent static response of each segment (see Lewis 1999; Winters 2002, and Mullins *et al.* 2002). However, it is not the intent of this paper to promote a particular analysis procedure or execute a complete analysis but to merely highlight the possible variations in deep foundation segmental load distribution as a result of modulus selection, or rather the selection of a particular stress-strain-strain rate relationship.

5   CONCLUSION

Long piles subjected to significant amounts of side shear are often constructed with embedded strain gages positioned at strategic locations throughout the pile so as to better distinguish the load contributing components (end bearing *vs.* side shear) and develop the side shear load distribution. In load tests that produce low concrete strains and/or strain rates, an approximated linear stress-strain relationship and constant modulus may prove to be sufficient for determining the stress at strain gage locations. Tests exhibiting low strain rates but high strains are best evaluated using a non-linear, parabolic stress strain relationship. However, in tests that produce both high strains and strain rates, the evaluation of the true concrete stress requires more sophisticated analysis. The presented approach incorporates the effects of strain magnitude, strain rate, and the rate of change of strain rate. As a result, the concept of a linear elastic modulus in concrete is virtually unusable.

Based on the results of the nonlinear hysteretic model developed herein, certain conclusions can be drawn:

- A linear stress-strain relationship (constant modulus) at best slightly over predicts stresses at upper gage levels and under predicts stresses at lower gage levels; in the worst case it misrepresents all gage levels.
- Since the use of ACI and modulus gage values are constant modulus assumptions, the segment t-z curves will show similar geometric shapes, but differ in value by the same degree as the difference in assumed moduli. But, neither accounts for the true nonlinearity or strain rate dependency.
- Since most foundations are not loaded to structural failure, a concrete stress-strain-strain rate relationship purely based on concrete break data is inadequate in describing load/unload cycles found in load tests.
- In all concrete specimens that undergo a load cycle, the stress follows a common surface up to the point of strain inflection or maximum strain rate (herein referred to as the transition point). However, each specimen returns along a different path defined by the strain rate at the transition point which can vary depending on the rate and magnitude of loading.

In order to implement the proposed model toward the regression of load test data, the following procedures are recommended:

- Concrete cylinder compression tests should be performed in accordance with ASTM C39 on specimens from each pile to identify the ultimate compressive strength (f_c'), the strain at ultimate strength (ε_o), and the strain rate at which the ultimate strength and strain were determined $(\dot{\varepsilon}_o)$. If analyzing static load test data, the Hognestad formula is sufficient.
- Strain gages should be placed at or near ground surface such that the measured load can be used to define the concrete stress-strain relationship prior to the occurrence of load shed. This measured load should be inertia-corrected and discounted depending on the reinforcement. Although it is ideal to place the gages above the ground surface, some piles and shafts do not extend above the ground. In these circumstances, the gages should be placed reasonably shallow.
- Using the values from the concrete cylinder compression tests, normalize the stress, strain, and strain rate (if necessary) at the upper gage level. Plot the results against the modeled response to determine whether the model sufficiently predicts the stress path.

- If the model sufficiently predicts the stress at the upper gage level, then apply the model to all gage levels.
- If performing a rapid load test, further regress the data to obtain the equivalent static response.

NOTATION

E_c = Young's modulus or elastic modulus;
f_c = stress at a particular strain;
f_c' = compressive strength or ultimate strength;
f_N = stress normalized with respect to compressive strength;
w_c = unit weight of concrete;
ε_c = concrete strain;
ε_N = strain normalized with respect to ε_o;
ε_o = concrete strain at f_c';
$\dot{\varepsilon}_c$ = instantaneous concrete strain rate;
$\dot{\varepsilon}_o$ = concrete strain rate normalized to $\dot{\varepsilon}_o$;
$\dot{\varepsilon}_{N-tp}$ = normalized concrete strain rate at the transition point; and
$\dot{\varepsilon}_o$ = concrete strain rate corresponding to ASTM loading rate.

REFERENCES

ASTM C 39-01. "Standard Test Method for Compressive Strength of Cylindrical Concrete Specimens." American Society for Testing and Materials, West Conshohocken, PA.

Fellenius, B. (2001). "From Strain Measurements to Load in an Instrumented Pile." Geotechnical News, March, 35–38.

Fu, H.C., Erki, M.A. and Seckin, M. (1991). "Review of Effects of Loading Rate on Reinforced Concrete," Journal of Structural Engineering, Vol. 117, No. 12, December, 3660–3679.

Lewis, C. (1999). Analysis of Axial Statnamic Testing by the Segmental Unloading Point Method, Master's Thesis, University of South Florida, Florida.

Mullins, G., Lewis, C. and Justason, M. (2002). "Advancements in Statnamic Data Regression Techniques." Deep Foundations 2002: An International Perspective on Theory, Design, Construction, and Performance, ASCE Geo Institute, GSP# 116, Vol. II, 915–930.

Mullins, G. and O'Neill, M. (2003). Pressure Grouting Drilled Shaft Tips: A Full-Scale Load Test Program, May.

Stokes, M. (2004). Laboratory Statnamic Testing, Master's Thesis, University of South Florida, Florida.

Takeda, J. and Tachikawa, H. (1971). "Deformation and Fracture of Concrete Subjected to Dynamic Load." Proceedings of International Conference on Mechanical Behavior of Materials. Kyoto, Japan, 267–77.

Delft Cluster project

Chapter 7

Rapid model pile load tests in the geotechnical centrifuge (I)

N.Q. Huy
Delft University of Technology, Delft, The Netherlands

A.F. van Tol
Delft University of Technology, Delft, The Netherlands
Department of Civil Engineering and Geosciences, Deltares, Delft, The Netherlands

P. Hölscher
Deltares, Delft, The Netherlands

SUMMARY

Experimental research on the influence of the loading rate and of generated pore water pressure during rapid load tests is carried out. In a Geotechnical centrifuge, a number of tests on piles in sand were performed. The influence of loading rate and drainage condition are the main item of this research. This paper describes the testing programme and the test set-up. The test data with typical measurement results from each load test are presented. Next the effects of the loading rate on the pile resistance are shown. The theoretical result that a rapid load test might overestimate the static capacity due to pore water pressure is confirmed. The results of the pore pressure measurements, the evaluation and conclusions are presented in the second paper on this topic hereafter.

I INTRODUCTION

Rapid pile load test methods such as a Statnamic test or a Pseudo-static test are considered as an efficient alternative method for static pile load testing because of its fast execution and relatively low costs. During the test, the pile is loaded with a loading of a duration from 50 ms to 200 ms, which is 10 to 20 times longer than a dynamic pile load test. Therefore, the stress wave phenomenon in the pile is negligible. However, for this loading duration, excess pore water pressure is generated in soil closes to the pile even if it is located in sandy soil (Hölscher 1995 and Maeda 1998). The excess pore water pressure may affect either the stiffness or the ultimate bearing capacity of the pile. This paper studies these effects based on experimental research by performing a number of rapid pile load tests in a centrifuge.

Small-scale model tests have long been used to study prototype behaviour in many geotechnical problems. Full-scale tests are costly, time-consuming, and are not always possible. In pile foundation engineering, model tests offer the opportunity to investigate many aspects of pile behaviour in conditions that can be controlled and reproduced (Sedran *et al.* 1998). Eiksund and Nordal (1996) measured excess pore pressure near the model pile tip in their experiments. The overall pore pressure response is as follows: a small increase is observed initially, but the excess pore pressure turns into a negative value. The study concludes that the excess pore pressure induced by pile penetration has a minor influence on resistance. These experiments were performed in a calibration chamber at a 1-g condition. This situation does not represent the initial stress condition correctly. Because the soil behaviour is highly non-linear and dependent on the stress level, soil behaviour in the chambers may not be the same in the prototype and findings from these experiments may also differ. As indicated by Altaee and Fellenius (1994), these test results are less relevant to the real behaviour of a pile-soil system in a full-scale prototype.

The geotechnical centrifuge can be used to overcome the limitation of 1-g devices. A model scaled using a factor N can be tested in the centrifuge, if centrifugal accelera- tion of N times higher than earth's gravity is applied to the sample. If so, the incre- ment in vertical stress per scaled length in the model equals the increment of stress per length in the prototype. The initial stress in the model and prototype is therefore identical, and soil behaviour in a small-scale model will be almost identical to that in a prototype.

A number of centrifuge experiments described in literature is relevant to the topic of rapid pile load testing in a centrifuge (Allard 1990; de Nicola and Randolph 1994; Bruno and Randolph, 1999). Their tests focus on the behaviour of piles or surround- ing sand during a dynamic pile load test, but none adequately considers the pore pres- sure response. Allard (1990) performed the experiments in dry sand. De Nicola and Randolph (1994) and Bruno and Randolph (1999) used silica flour instead of sand to reduce the permeability of the soil sample. Their paper does not mention the effect of excess pore pressure.

The response of pore pressure may significantly affect a pile's mobilised resistance during a load test, and the effects depend on the sand's drainage condition and the generation of pore water pressure during a test. To obtain more knowledge about mobilised pile resistance during a rapid load test it was therefore decided to carry out a series of pile load test in the geotechnical centrifuge at Deltares (GeoDelft), giving particular attention to modelling the pore pressure response. The test series and re- sults are presented and discussed in this paper.

The test series included a number of axial load tests on a model pile founded in a well-defined saturated sand bed. The loading rate of these load tests was varied. Simu- lation of a prototype static pile load test was the slowest load test, the fastest was a simulation of a prototype rapid pile load test. Some intermediate loading rates were tested as well. The principle aims of the test series were:

1   To study the effect of the penetration rate on the resistance of a pile embedded in sand.
2   To obtain knowledge about excess pore pressure in the soil around the pile tip, and its effect on resistance during a rapid pile load test.
3   To validate numerical results concerning the effects of excess pore pressure on pile resistance, presented in Huy et al. (2007).

First the testing programme and the scaling rules are presented. Next the test set-up is described and discussed. Test data with typical measurement results from each load test are presented next, as well as the effects of the loading rate on the pile resistance. The results of the pore pressure measurements, the evaluation and conclusions are presented in the second paper on this topic hereafter.

2   SCALING RULES

The standard scaling rules for centrifuge modelling have been well established in literature (e.g. Altaee and Fellenius 1994; Sedran *et al.* 2001; Garnier *et al.* 2007) and will not be repeated here. This section first introduces a set of scaling rules, which can be used to extrapolate the results from the test series into an equivalent prototype situation. Secondly, the main point of discussion is how to deal with the pore fluid to correctly model the pore pressure response during a prototype rapid load test.

Centrifuge modelling is used to obtain identical stresses and strains in the model as in the prototype. Generally, if the scaling factor N is chosen for the length the acceleration level in the centrifuge model will be N times higher than in the prototype. Other quantities can be found from the dimensional analysis. The static and rapid pile load tests simulated in this test series were not the reduced scale model of a specific prototype case. The scaling rules introduced here are used to extrapolate the test results into an equivalent prototype situation. Table 1 presents the scaling rules.

The requirements for pore fluid treatments are considered next. When considering the permeability of a soil sample as defined by Darcy's law $k = Kg/v$ (where K is the intrinsic permeability of the sand, g is the acceleration level, and v is the viscosity of the pore fluid), it can be seen that permeability depends on the acceleration level. In the centrifuge environment, acceleration is increased N times as well as soil permeability. This implies that if the sand and water in the centrifuge and prototype are the same, the pore pressure dissipation process (consolidation) in the centrifuge will occur N^2 times faster. This conflicts with the time scale presented in Table 1. To compensate for this and to retain the same sand, a fluid with a viscosity N times higher than water should be used, as proposed by Fuglsang and Ovesen (1986). However, the viscosity of pore fluid used in this test series will differ somewhat from the requirement necessary to meet the aims of this study. This is discussed in the following paragraphs.

Table 1 The scaling rules.

Parameter	Model	Prototype
Length and displacement	1	N
Area	1	N^2
Volume	1	N^3
Time	1	N
Acceleration	N	1
Velocity	1	1
Density of soil	1	1
Mass	1	N^3
Force	1	N^2
Stress	1	1
Strain	1	1

Huy *et al.* (2007) have indicated that the significance of the effect of excess pore pressure on the pile resistance in a rapid load test depends on the drainage condition of the soil, which is determined by the defined dynamic drainage factor η:

$$\eta = \frac{GT}{g\rho R^2} k = \frac{GT}{\rho R^2} \frac{K}{v} \tag{1}$$

where: G is shear modulus, T is the duration of the loading, ρ is soil density, R is pile radius.

In order to simulate the generation and dissipation of excess pore pressure in the centrifuge tests as realistic as possible the starting point was to keep the drainage factor in the model and the prototype equal. The ratio of the drainage factor in the model and in prototype depends on the relative value of the permeability. If water is used in the centrifuge tests, the drainage factor will be N times smaller than in the prototype, since time scales with $1/N$. If a fluid with N times higher viscosity is used, the drainage factor will be identical.

The scaling factor $N = 40$ is chosen in the test series for the convenience of the test performance.

In the planning phase, the goal was to study the effects of pore pressure in the relevant range of the drainage factor. However, the hydraulic actuator (plunger) was not fast enough to reach the low values of drainage factor (=0.01). Taking the loading duration of the Statnamic test (100 milliseconds) as a representative loading duration for a rapid pile load test in prototype, the loading duration of the model test should be 2.5 milliseconds with a scaling factor of 40. In reality, the fastest loading duration of the plunger was approximately 7.5 milliseconds, thus 3 times slower than the requirement. To compensate for this, the viscosity of the pore fluid had to be increased another three times (i.e. $3 \times 40 = 120$ times higher than water). With that increment however, the drainage factor η in the fastest rapid test was close to 1, still too high to show clear effects of excess pore pressure according to Figure 1, (Huy *et al.* 2007), where F_{PD} and F_D are the ultimate resistance in partially and fully drained conditions. It was therefore decided to increase the viscosity of the fluid to lower the drainage factor. In centrifuge tests 2 and 3, the viscosity of the fluid was chosen as approximately 300 times higher than the viscosity of water. In

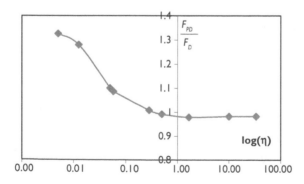

Figure 1 Ultimate toe resistance as a function of the dynamic drainage factor, (Huy *et al.* 2007).

Table 2 Drainage factor in the model and prototype.

	Test No.	Id (%)	G (MPa)	T_model (s)	viscosity cP	k0_water (m/s)	ko_fluid (m/s)	η_model (–)	η_prototype (–)
Test 2	RLT2-1-0.1D	54	39.51	0.0529	265	8.85E-04	3.34E-07	2.19	14.49
	RLT2-2-0.1D	54	39.51	0.0203	265	8.85E-05	3.34E-07	0.84	5.56
	RLT2-3-0.1D	54	39.51	0.0100	265	8.85E-05	3.34E-07	0.41	2.74
Test 3	RLT3-1-0.1D	36	20.68	0.0529	292	8.85E-05	3.03E-07	1.04	7.58
	RLT3-2-0.1D	36	20.68	0.0203	292	8.85E-05	3.03E-07	0.40	2.91
	RLT3-3-0.1D	36	20.68	0.0100	292	8.85E-05	3.03E-07	0.20	1.43
Test 4	RLT4-1-0.1D	65	51.04	0.0529	1	8.85E-05	8.85E-05	748.52	18.71
	RLT4-2-0.1D	65	51.04	0.0203	1	8.85E-05	8.85E-05	287.24	7.18
	RLT4-3-0.1D	65	51.04	0.0100	1	8.85E-05	8.85E-05	141.50	3.54

test 4, it was decided to use water as a pore fluid to achieve a nearly fully drained drainage condition. In this way, centrifuge tests 2 and 3 can be seen to simulate a prototype rapid load test on a pile founded in a Baskarp sand with a permeability approximately 2.5 times lower (300/120 = 2.5). For centrifuge test 4 the permeability is 40 times higher. Table 2 presents the drainage factor values for every RLT with an imposed displacement of $0.1*D$. The values in the column 'η_model' are the actual values in the model scale, and the values in the column 'η_prototype' are the values for an equivalent prototype RLT with a loading duration ($T = 40*T$_model), performed on a pile in Baskarp sand.

The viscous fluid chosen for this test series was developed at Delft Geotechnics (Allard and Schenkeveld 1994). It is a mixture of water and sodium carboxy Methyl Cellulose. The viscous fluid can reach a viscosity up to 300 times the viscosity of water, while its physical properties are similar to those of water. Extensive laboratory tests have shown the similarity in constitutive behaviour between a sand specimen saturated with viscous fluid and a sand specimen saturated with water (Allard and Schenkeveld 1994).

3   EXPERIMENTAL SET-UP

This section describes components of the test set-up, preparation of the sand bed, and the instrumentation used in the test. The effects that the set-up may have on test results are considered at the end of the section.

3.1   Test set-up

Figures 2 and 3 show the test set-up. The load tests were carried out in a Ø600 mm steel sand fill container. A loading frame with plungers was mounted above the container. The model pile was connected to the plungers. The most important components of the set-up will be described briefly in the following paragraphs.

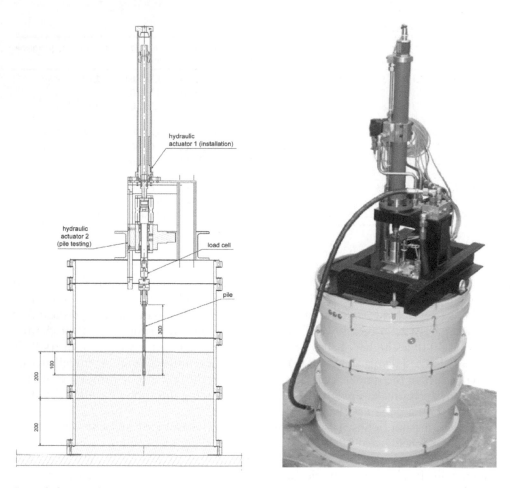

Figure 2 Sketch of centrifuge test set-up. *Figure 3* Photograph of centrifuge test set-up.

Loading system

The loading system consisted of two hydraulic actuators (plungers) that were connected in series. The first and largest plunger was fixed on the loading frame, and was used to install the pile to its starting point before the load tests began. The second smaller plunger was the fast loading plunger, and was fixed to the rod of the first plunger. This second plunger was used to perform the model pile load tests. The pile was attached to the second plunger. The loading frame was connected to the top of the container.

The model pile

The model pile was a steel pile with a length of 300 mm, a diameter of 11.3 mm and a weigh of 570 grams. A load cell was placed on the model tip to measure pile tip resistance. The pile tip was also equipped with a pore pressure transducer to measure pore pressure directly below the pile tip. For this purpose, a small hole, diameter 5 mm was made to accommodate the transducer. A photograph of the model pile is shown in Figure 4.

3.2   Sample preparation

Soil properties

Baskarp sand with a $d_{50} = 130$ μm was used for the tests. It is widely used for laboratory tests, and its soil parameters have been reported in a variety of literature (e.g. Allard *et al.* 1994; Mangal 1999). The grain size distribution of the sand used is shown in Figure 5. The sand's basic soil parameters, determined from the GeoDelft laboratory test, are presented in Table 3.

Sample preparation

The following steps were followed to prepare a homogenous sand body at a predetermined density. The container was first filled with de-aerated water and the

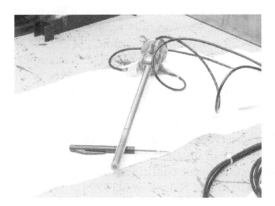

Figure 4 The model pile.

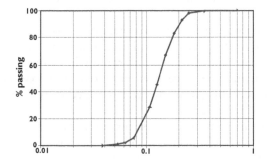

Figure 5 Grain size distribution of Baskarp sand.

Table 3 Properties of Baskarp sand.

Parameter	Value	Dimension
Density grains	2,647	kg/m^3
d_{10}	90	μm
d_{50}	130	μm
d_{90}	200	μm
Min. porosity	34	%
Max. porosity	46.9	%
Permeability at min. porosity	6.5*10^{-5}	m/s
Friction angle at $R_D = 50\%$ ($n = 40\%$)	41	Degr.

pre-determined amount of wet sand was pluviated under the water surface. The water level in the cylinder was sufficient and the sand was softly blown into the water in an upwards direction to slow the grains to the equilibrium speed to create a very loose sample. The loose sand sample was then compacted. A loaded permeable plate was placed on the surface of the sand sample, the complete container was lifted a few centimeters above the floor, and the container was released. The impact caused by falling compacted the sand sample. By repeating the process and carefully registering the achieved height, the predetermined relative density (D_r) could be achieved. When the desired density was reached, the top layer was carefully removed and flattened. This preparation method, described by Van der Poel and Schenkeveld (1998) allows to prepare soil samples with a predefined relative density within 1–2% accuracy.

In those cases where viscous fluid was used, the viscous fluid replaced the saturated water in the prepared sand sample. The viscous fluid was carefully positioned above the saturated sand sample. A vacuum was then applied at the bottom of the container and the viscous fluid penetrated into the sand sample. As the colour of the viscous fluid was purple, a colour change was observed in the drainage pipe once it reached the bottom of the sand sample. The viscosity of the fluid was measured, and the saturation process was stopped when the measurement confirmed the value of viscosity. This saturation process is described in Allard and Schenkeveld (1994).

3.3   Measurement set-ups

The following parameters were measured as a function of time during each test:

- V_PL_KL: Displacement of the small plunger. This was measured using a transducer, which was an integral part of the servo-control system and was used to obtain the load-displacement characteristics in static and rapid pile loading.
- F_PL_BK: The load on the pile head. The force was measured by a load cell, which was mounted at the pile head. This parameter was required to obtain the load-displacement characteristics of the pile head.
- F_PL_OK: The force on the pile tip. The force was measured by a specially-constructed flat cone tip equipped with a load cell. This parameter was required to obtain the load-displacement characteristics of the pile tip.

–   WSM_PL: Pore pressure beneath the pile tip. The pore pressure immediately under the pile tip was measured using a pore pressure transducer integrated into the pile tip.
–   WSM 1–4: Pore pressure in the sand bed. The pore pressure was measured at four different positions beneath the pile tip level. The transducers provided information about pore pressure variation as a function of time during a load test. Figure 6a and 6b shows the location of the four pore pressure transducers with respect to the pile tip location at a depth of 20D (226 mm) below the sand surface and their installation in the container.

All the measurement devices were calibrated before each centrifuge test. The load cell was a miniature force transducer U9B made by HBM Inc., with a measurement range of 0–10 kN. The calibration in this range showed a maximum absolute fault less than 0.007 kN. The pore pressure transducers were high performance pressure transducers produced by Druck Ltd. The pore pressure transducer in the pile tip was a PDCR 42 Druck type, with a maximum range of 2000 kPa. The calibration up to 1000 kPa showed the maximum fault to be less than 0.3 kPa. The pore pressure transducers in the sand bed were a PDCR 82 Druck type, with a maximum range of 1000 kPa. It was calibrated up to 700 kPa and showed the maximum fault to be less than 0.26 kPa.

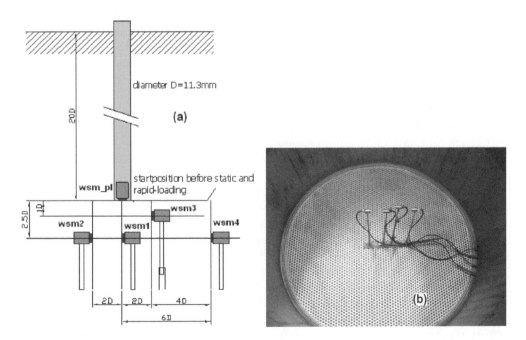

Figure 6 (a) Positions of pore pressure transducers; (b) installation.

3.4   Discussion of the set-up effects

Some aspects of the model test set up may have an undesirable effect on the test results. For the described set-up, the following items are discussed below:

- – Grain size of the sand
- – Size of the sand fill container
- – Size of the pore pressure transducers
- – Soil condition changes due to installation and the load tests
- – Difference between the model load test and the prototype rapid load test.

The grain size used for the centrifuge modelling corresponds to those N times larger in the prototype. This may induce differences in behaviour between the model and the prototype, due to the dependence of the failure mechanism (defined in terms of the width and extent of the shear band) on the grain size. However, scaling the sand grain size in the centrifuge model would require using the particle size of silt or clay, which have significantly different strength characteristics. In most cases, the same soil is therefore used in the model and in the prototype. Ovesen (1981) performed a series of pull-out tests on model anchors where the ratio of the anchor diameter to the mean grain size of sand was from 25 to 128. No effect was found on grain size. Yamaguchi *et al.* (1977) studied the effect on the bearing capacity of the footing. The ratio of the footing diameter to the mean grain size was varied from 36 to 286, but no effect was observed. Phillips and Valsangkar (1987) performed a series of centrifuge tests to study the grain size effects on penetration resistance. In their tests, the acceleration level was from 20 g to 80 g, and the ratio between the probe diameter and mean grain size was from 10 to 58. They found that the effect of grain size is observed if the ratio of probe diameter to mean sand grain size is smaller than 20. In this study, the mean grain size of Baskarp sand D_{50} was 0.13 mm (Table 3) and the model pile diameter was 11.3 mm, thus a ratio of more than 86. Therefore grain size effects are not expected in the centrifuge tests.

In the model test, the soil sample volume was limited by the size of container. This differed from the prototype situation, where the soil was an infinite half-space. The difference may have caused unexpected boundary effects on the test results. The significance of these effects depends on the size of container, the diameter of the penetration object, and the type of test carried out. For these model tests, the major boundary effects were: (1) effect of the container size to resistance of the model pile; (2) effect of the reflection of stress waves from the container walls.

The first effect generally exists when a penetration test is carried out in a narrow container, but it can be negligible at the applied ratio between the container diameter and the model pile diameter (diameter ratio). From their calibration chamber tests, Parkin and Lunne (1982) showed that the effect is more pronounced as the density of the soil sample increases. They suggested that the diameter ratio between a container diameter and a model pile diameter should not be below 50 to eliminate the effect at all densities. Gui (1995) suggested that a diameter ratio value of 40 is enough to ignore the boundary effect, based on his centrifuge test on silica Fontainbleau sand. Other authors have also suggested different values for the diameter ratio, but the value of 50 is generally accepted. In this test, the container's inner diameter was 589 mm and the pile diameter was 11.3 mm, which makes the diameter ratio larger than 52. The container size in this case will therefore have a negligible effect on the resistance of the model pile.

The second effect is related to a dynamic model test where stress waves are generated. These waves radiate from the model pile into the sand sample and reflect back from the container walls. For the container used in this test, the effect may be more significant as the container was circular. This facilitated the reflection, and focused the reflected waves to the centre of the container where the model pile was placed. Solutions to reduce the effect include using a type of wave-absorbent material along the container walls, or enlarging the size of the container. However, wave reflection in this case is not expected to be significant due to the type of tests performed. Wave reflection is known to be more important for low amplitude vibration tests, while in this case of pile load testing there was large plastic deformation and sliding between pile and soil, which reduced the importance of (elastic) wave transmission from the pile into the surrounding soil. The eventual existence of wave reflection during the RLTs, would be seen in the signal of the pore pressure transducers. Later descriptions will show that there was no evidence of wave reflection for a RLT with a penetration rate up to 80 mm/s. It only seems to be observed in the fastest tests (v≈300 mm/s), but the tests results are the similar to the slower RLTs. It can therefore be concluded that the reflected waves are small and their influence is negligible in the test set-up.

4   TEST PROGRAMME

The test series included four centrifuge tests that differed in the initial density of the sand sample and the viscosity of the pore fluid. The first pilot test is not considered here, since major changes in the test set-up have been introduced since that test. Table 4 provides an overview of conditions during the remaining three tests, in both the geo-centrifuge situation as well as the equivalent prototype situation.

A number of static load tests (SLT) and rapid load tests (RLT) were performed in each centrifuge test. Figure 7 shows the sequence of load tests performed during each centrifuge test. For the sake of convenience, the term 'centrifuge test' is referred to as 'the test', and 'load test' means a particular pile load test (step in Figure 7) performed in a centrifuge test.

During preparation, the pile tip was pre-embedded at a depth equivalent to 10 times the pile diameter (10*D) below the surface. The load test sequence shown in Figure 7 was

Table 4 Overview of the centrifuge tests.

Parameters	Test 2	Test 3	Test 4
Relative density	54%	36%	65%
Material	sand	sand	sand
Pore fluid	viscous fluid	viscous fluid	water
Viscosity	265	292	1
Prototype			
Material	sand	sand	sand
Permeability	fine sand	fine sand	coarse sand

step	Test	RLT1 slow	RLT2 medium	RLT3 fast
0	Installation			
1	SLTi-1			
2	RLTi-1 - 0.01D			
3	RLTi-1 - 0.02D			
4	RLTi-1 - 0.05D			
5	RLTi-1 - 0.1D			
6	SLTi-2			
7	RLTi-2 - 0.01D			
8	RLTi-2 - 0.02D			
9	RLTi-2 - 0.05D			
10	RLTi-2 - 0.1D			
11	SLTi-3			
12	RLTi-3 - 0.01D			
13	RLTi-3 - 0.02D			
14	RLTi-3 - 0.05D			
15	RLTi-3 - 0.1D			
16	SLTi-4			
i = 2, 3, 4 is the centrifuge test number				

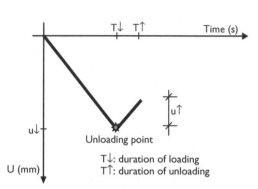

Figure 7 Loading scheme of the tests.

Figure 8 Displacement pattern of a rapid load test.

applied after the centrifuge had been accelerated to a level of 40 g. The test began with installation of the model pile. The first hydraulic actuator pushed the model pile deeper into the sand bed with a velocity of 10 mm/min and a total displacement of 11.3 cm (10*D). Once installation was complete, the pile tip reached a depth of 20*D and the first static load test (SLT) was carried out. Three sets of four rapid load tests (RLT) were then performed, followed by another SLT. The SLT was performed with a velocity of 0.00167 mm/s and a displacement of 10% of the pile diameter (0.1 * D). The duration of the load was shorter for each set of four RLTs, which led to an increasing test loading velocity. Each of the four RLTs was carried out with increasing maximum displacement (1%, 2%, 5% and 10% of the diameter). All load tests were displacement-controlled with the displacement pattern shown in Figure 8. Extra RLTs were performed in tests 3 and 4, using the same imposed pile head displacement (0.1D) but different penetration rates.

5 TEST RESULTS

This section presents an overview of the results from the model pile load test regarding the measured forces and motions of the model pile during the load tests. The second paper, hereafter, will show the measured response of pore pressure during rapid load tests. All results are presented in the model scale.

The following terms and parameters will be used.

– Pile head force (F_head) is a directly measured parameter.
– Pile tip force (F_tip) is also a directly measured parameter.
– Shaft force (F_shaft) is derived from the difference between F_head and F_tip.
– The name used for a static load test will be SLTi-1, which is a first static load test (step 1 in Figure 7) of the centrifuge test i (i = 2, 3, 4).

– The name used for a rapid load test will be in the form of RLTi–3–0.02D, which is the third rapid load test with the imposed pile head displacement of 2% of pile diameter (step 13 in Figure 7) of the centrifuge test *i*.

5.1 Static Load Test (SLT)

The typical example of the measured forces and displacement of the model pile in a static load test are shown in Figure 9 and the related force-displacement curves in Figure 10. The results are taken from the first static load test in test 3, where viscous fluid was used as the pore fluid and the initial relative density of the sand bed was 35%. The SLT was performed with a velocity of 0.00167 mm/s and a displacement of 10% of pile diameter (0.1*D). Figure 10 shows that most of the mobilised soil resistance of the model pile is end-bearing resistance, which is approximately 80% of the pile's total capacity. The maximum shaft resistance reaches an ultimate value at a relatively small displacement of the model pile (approximately 3% of the pile diameter), whereas the ultimate tip resistance requires larger displacement (about 7% of the pile diameter).

Four SLTs were carried out in each centrifuge test. A comparison between them gives information about the change in soil condition due to RLTs performed in between. All the force-displacement curves of the SLTs in centrifuge tests 2, 3, and 4 are shown in Figures 11, 12, and 13 respectively. Some deviation can be seen between the curves in each figure, and the later SLT shows generally higher resistance than the previous SLT. Because the four RLTs were performed between two consecutive SLTs, densification of the sand due to these RLTs can be expected, resulting in the observed deviation of the SLTs. This explanation is strengthened by the fact that the largest deviation is observed in test 3, with the lowest initial density ($I_D = 35\%$), thus the highest densification to be expected. Also, the deviation in test 2 ($I_D = 53\%$) is less, and similar deviation is observed in test 4 with the highest initial density ($I_D = 65\%$). An exception is seen with the curve known as SLT4-1 in Figure 13. The reason is not clear. Nevertheless, this exception does not affect the results to much since the comparisons are made between the resistance of a RLT and its closest SLT.

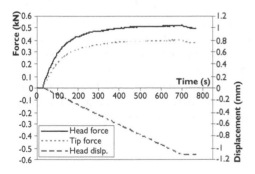
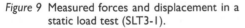

Figure 9 Measured forces and displacement in a static load test (SLT3-1).

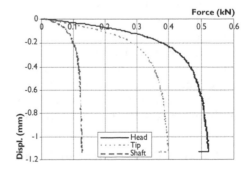

Figure 10 Force-displacement curves (SLT3-1).

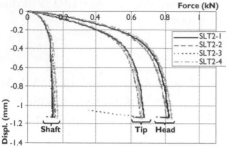

Figure 11 Force-displacement curves of the SLTs in test 2.

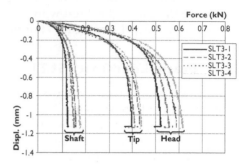

Figure 12 Force-displacement curves of the SLTs in test 3.

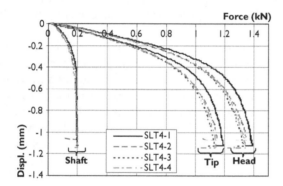

Figure 13 Force-displacement curves of the SLTs in test 4.

5.2   Rapid Load Tests (RLT)

This section presents the RLT measurement data. During one centrifuge test, twelve RLTs were carried out with different imposed displacements and penetration rates. Data measured during the RLTs are generally very similar in all cases of corresponding pile head displacement. From the RLTs with the same displacement, only the data of one, representative RLT are presented here. The representation includes the measured forces, displacement during a RLT, pile velocity and acceleration derived from the measured displacement, and the load-displacement curves. It shows that RLT results with a penetration rate less than 0.08 m/s are not affected by the inertial force (i.e. the inertial force is very small in comparison with the pile head force) and that tests with higher rates are very similar to the prototype rapid load test. The effect of different pore fluid usage on the RLT data is discussed in paper (2).

RLT with a displacement of 1% pile diameter

Typical measured parameters during a RLT with an imposed pile head displacement of 0.113 mm (0.01*D) as a function of time are shown in Figure 14. These are the

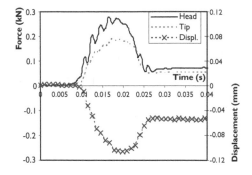

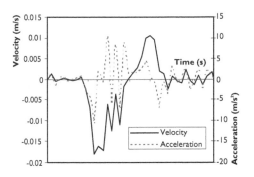

Figure 14 Typical measured parameters in a RLT with a displacement of 0.01*D.

Figure 15 Typical derived velocity and acceleration in a RLT with a displacement of 0.01*D.

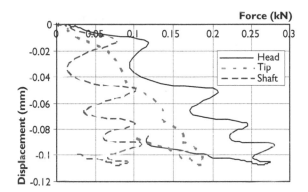

Figure 16 Typical load-displacement curves of a RLT with a displacement of 0.01*D.

results of test RLT3-1-0.01D where the imposed velocity was ~0.012 m/s. The pile head velocity and acceleration shown in Figure 15 are derived from the measured displacement. The pile velocity varies somewhat during the loading because the loading plunger displacement is to small at this loading velocity. The acceleration of the pile is in the magnitude of 10 m/s^2, thus the magnitude of inertial force (6 N, with a pile mass of 0.57 kg) is negligible to the maximum pile head force (250 kN). The force-displacement curves are shown in Figure 16. The resistance of the model pile does not reach the ultimate value due to the small imposed displacement.

RLT with a displacement of 0.02 pile diameter (RLT03-3-0.02D)

Typical results of a RLT with an imposed pile head displacement of 0.02 × D are shown in Figures 17, 18, and 19. These results are from the test RLT03-3-0.02D where the imposed penetration rate was 0.0251 m/s. In general, the results are similar to those of the RLT with a displacement 0.01*D. The pile head displacement is still too small to fully mobilise the soil resistance.

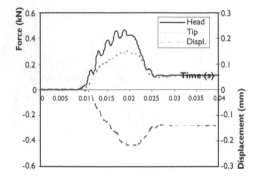
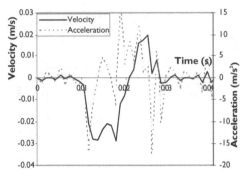

Figure 17 Typical measured parameters in a RLT with a displacement of 0.02*D.

Figure 18 Typical derived velocity and acceleration in a RLT with a displacement of 0.02*D.

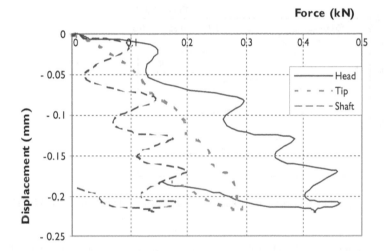

Figure 19 Typical load-displacement curves of a RLT with a displacement of 0.02*D.

RLT with a displacement of 0.05 pile diameter (RLT02-2-0.05D)

Typical results of a RLT with a pile head displacement of 0.05*D are shown in Figures 20, 21, and 22. These results are from the test RLT02-2-0.05D where the imposed penetration rate was 0.0305 m/s. These Figures show that the displacement and velocity patterns are very similar to those prescribed. Comparing the displacement patterns of the RLTs with imposed displacement of 0.01*D and 0.02*D show that the loading plunger requires a displacement of order 0.5 mm to perform properly. The shaft resistance reaches its ultimate value during the test, but displacement is not high enough to reach the ultimate toe resistance.

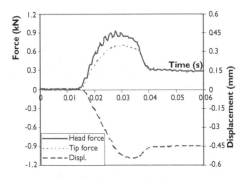

Figure 20 Typical measured parameters in a RLT with a displacement of 0.05*D.

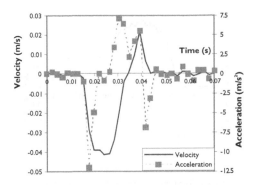

Figure 21 Typical derived velocity and acceleration in a RLT with a displacement of 0.05*D.

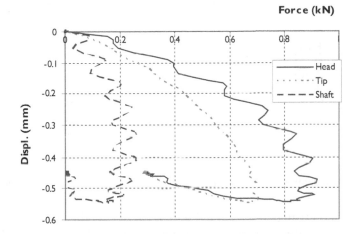

Figure 22 Typical load-displacement curves of a RLT with a displacement of 0.05*D.

RLT with a displacement of 0.1 pile diameter, velocity about 0.06 m/s

Typical results of a RLT with a pile head displacement of 0.1*D and pile velocity about 0.06 m/s are shown in Figures 23, 24, and 25. These results are from the load test RLT2-2-0.1D with an imposed penetration rate of 0.061 m/s. The actual velocity is seen to be somewhat higher than proposed, but the pattern is as expected. As shown in Figure 23, the displacement pattern is similar to the proposed pattern presented in Figure 20. This implies that RLT performance is well-controlled up to this penetration rate, and that the displacement and velocity patterns are as expected. Acceleration is in the magnitude of 20–40 m/s², hence the inertial force is still small in comparison with the pile head force. The shaft resistance reaches its ultimate value during the test. The ultimate value of tip resistance does not seem to be reached completely as no clear failure mode is seen. This type of pile tip resistance-displacement behaviour is

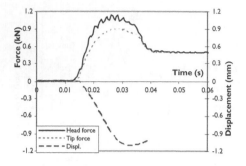

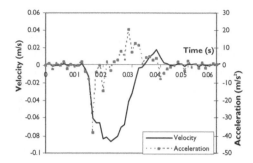

Figure 23 Typical measured parameters in a RLT with a displacement of 0.1*D; velocity < 0.08 m/s.

Figure 24 Typical derived velocity and acceleration in a RLT with a displacement of 0.1*D; velocity < 0.08 m/s.

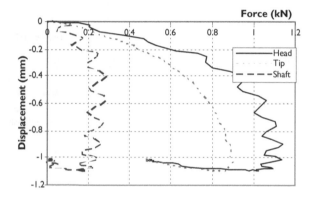

Figure 25 Typical load-displacement curves of a RLT with a displacement of 0.1*D; velocity < 0.08 m/s.

only observed in centrifuge tests 2 and 3 where viscous fluid is used as the pore fluid. It is not seen in centrifuge test 4 where water is used as the pore fluid. This will be discussed in part 2.

RLT with a displacement of 0.1 pile diameter, velocity higher than 0.1 m/s

This RLT was intended to have a penetration rate of 125.6 mm/s and a loading duration of 9 ms. However, after revising the test data it could be seen that the actual velocity was much higher (≈ 300 mm/s) and the loading duration was shorter (≈ 7 ms). Technical examination of the loading system showed that the loading plunger did not act in accordance with the proposed loading velocity. This probably exerted an impact load on the pile head at the highest possible velocity. Evidence of impact includes vibrations in the pile head force signal, and rebound in the pile head displacement. It is also manifested by the fact that the actual penetration rate and displacement pattern of this RLT are markedly different from those proposed, as shown in the next paragraph.

Figure 26 presents typical measured forces and displacement during the RLT with high velocity as a function of time. The pile head force shows oscillations and the

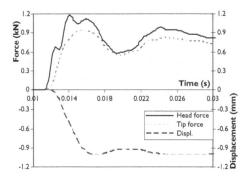

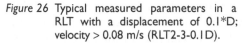

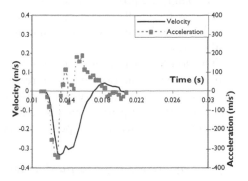

Figure 26 Typical measured parameters in a RLT with a displacement of 0.1*D; velocity > 0.08 m/s (RLT2-3-0.1D).

Figure 27 Typical derived velocity and acceleration in a RLT with a displacement of 0.1*D; velocity > 0.08 m/s. (RLT2-3-0.1D).

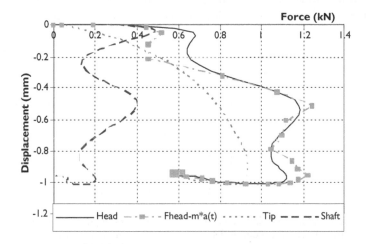

Figure 28 Typical load-displacement curves of a RLT with a displacement of 0.1*D; velocity > 0.08 m/s (RLT2-3-0.1D).

rebound of the loading plunger on the pile head. The maximum pile head displacement is approximately 1 mm, which is smaller than proposed. These observations denote that this RLT was out of control. However, the behaviour of the pile tip force seems to be very realistic (see dotted line in Figure 26). Figure 27 shows the values of velocity and acceleration as a function of time. A maximum velocity as high as 0.3 m/s is reached at the steady state of pile penetration. This value is in the same magnitude of pile velocity as during a prototype rapid load test (≈ 0.5 m/s). In contrast to other slower load tests, the acceleration of this RLT is relatively high and the inertial force is comparable with the pile head force at the start of the test (see Figure 28). The loading duration of the test is approximately 7 ms (equal to the loading duration of 40*7 = 280 ms in the prototype scale), which results in a test wave number (wave length/pile length) of approximately 120. This falls within the usual range of the rapid pile load test, according to the classification criterion proposed by the Japanese research group on rapid pile load test method (Kusakabe *et al.* 1998). Therefore, instead of all the

RLTs being the proposed constant rate of penetration load test (displacement control test), this fastest RLT was in fact closer to an impact load test (force control test). It can be regarded as a reduced scale of the prototype rapid load test in the equivalent soil condition.

The load-displacement curves of the RLT are presented in Figure 28. There is extreme over-valuation in shaft resistance at the start, which is caused by vibrations in the pile head force signal and an incorrect value for pile shaft resistance. The development of tip resistance is realistic and similar to the results of a slower RLT (e.g. Figure 25), where test performance was well-controlled.

Effect of different pore fluid usage

RLT data presented earlier are taken from centrifuge tests 2 and 3, where viscous fluid was used as the pore fluid. This section shows that the use of water as a pore fluid in test 4 does not affect the general behaviour of the model pile during a RLT, but does affect the failure mode in the tip force-displacement curve. The results of RLT4-3-0.1D are introduced in Figures 29, 30, and 31. This RLT was performed in the same

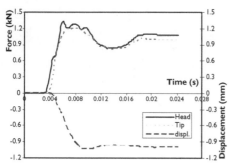

Figure 29 Example of measured parameters in test 4 (water as pore fluid; RLT4-3-0.1D).

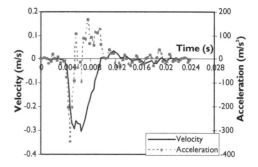

Figure 30 Example of derived velocity and acceleration in test 4 (water as pore fluid; RLT4-3-0.1D).

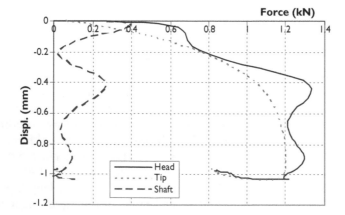

Figure 31 Example of load-displacement curves in test 4 (water as pore fluid; RLT4-3-0.1D).

way as RLT2-3-0.1D (Figures 26 to 28). Their results can therefore be compared to assess the effects. The general behaviour of the model pile shown in Figures 29 to 31 is very similar to that shown in Figures 26 to 28, apart from the pile tip force-displacement response. The tip force-displacement curve in Figure 31 clearly shows fully-plugged failure mode, which is not seen in Figure 28. This difference may be induced by the difference in excess pore pressure between these two RLTs, caused by different pore fluid usage as pointed out later in section 6.2.

6   EFFECT OF THE PENETRATION RATE ON PILE RESISTANCE

The effect of the higher penetration rate during a RLT on the measured pile resistance will be discussed. Since the ultimate value of pile resistance is the most important parameter, only load tests with an imposed displacement of 0.1D are considered here. The effect of penetration rate on shaft resistance and tip resistance is considered separately. It will be shown that the penetration rate affects both shaft resistance and tip resistance. This effect includes both, the rate effect (i.e. the increase in shear strength of the sand due to high shearing rate) and the effect of the excess pore pressure.

6.1   Effect of penetration rate on shaft resistance

Figures 32 to 34 show the total shaft resistance-displacement curves of the static load test and rapid load tests in centrifuge tests 2, 3, and 4 respectively. Each graph shows the static load-displacement curve and the curves for the slow, medium and fast tests. The total shaft resistance is derived from the difference between the measured pile head force and the pile tip force. The curves for the most rapid rate tests (penetration rate of approximately 280 mm/s) show extreme overshoot at the beginning, which must have been caused by vibration in the pile head force signal. This does not reflect the effect of the penetration rate, and the peak is therefore not considered further. The averaged values are evaluated.

Figures 32 and 33 show that total shaft resistance of the SLT is lower than that from the RLTs. What is more, shaft resistance is seen to increase as the RLT becomes

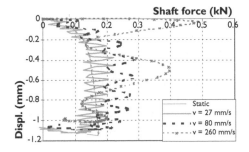

Figure 32 Shaft resistance-displacement curves (centrifuge test 2).

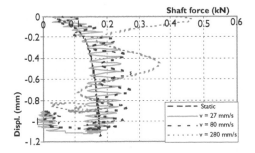

Figure 33 Shaft resistance-displacement curves (centrifuge test 3).

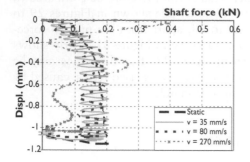

Figure 34 Shaft resistance-displacement curves (centrifuge test 4).

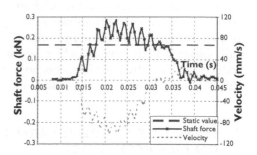

Figure 35 Pile shaft force and velocity vs. time (test 2, v = 80 mm/s).

faster. On the other hand, there is almost no difference between the curves shown in Figure 34, i.e. no penetration rate effect. As the range of penetration rates of load tests shown in Figures 32, 33, and 34 is the same, the observations must be related to a difference in test conditions between the centrifuge tests: the density of the sand sample and the pore fluid. Both can affect pile shaft resistance. However, the difference in sand bed density does not play a role in this case. Based on the results from the fast triaxial test series (Huy et al, 2005), the dense sand will show the highest effect. Figure 34, which shows results from centrifuge test 4 with the highest sand sample density ($I_D = 65\%$), shows no rate effects. If the effect does not exist in the test with the highest density (test 4, density = 65%), it does not exist in the tests with lower density (tests 2 and 3, density = 54% and 36%, respectively). This means that the increment in shaft resistance, which is observed in the tests 2 and 3 (shown in Figures 32 and 33), is solely relate to the influence of viscous fluid. Although the pore pressure is not measured at the pile shaft, it can be expected that the generation of negative excess pore pressure at the pile shaft will increase the effective stress and shaft resistance. Using this assumption, the difference between Figures 32, 33, and 34 can be explained by the difference in drainage ability. In test 4, water is used as the pore fluid (high permeability) and excess pore pressure along the pile shaft is low. Shaft resistance is therefore not affected. By contrast in tests 2 and 3, where viscous fluid is used as a pore fluid (i.e. the permeability is much lower), excess pore pressure would be significant enough to affect shaft resistance. As the drainage conditions in tests 2 and 3 are nearly the same, the penetration rate effects shown in Figures 32 and 33 are very similar.

Figure 35 presents the shaft resistance and pile velocity as a function of time for the test named "$v = 80$ mm/s" in Figure 32 and the static values of shaft resistance. Since the increment of shaft resistance is due to the negative excess pore pressure, values higher than the static value can only be observed at a certain time during the velocity increase phase where excess pore pressure increases. As pile velocity decreases, excess pore pressure starts to dissipate and shaft resistance therefore gradually decreases to the static value. Close to the point of maximum pile displacement, velocity reaches zero and the shaft resistance is virtually the same as the static value. At that moment the shaft resistance is not affected by excess pore pressure and equals the static value at the point of maximum pile displacement.

To summarise, in the case of a pile founded in sand as described here, shaft resistance is not affected by the rate effect but is affected by excess pore pressure. Sand permeability is the determining parameter. In the tests, the observed increment of the maximum shaft resistance did not affect the value of the pile resistance at the time of maximum displacement (unloading point).

6.2   Effect of penetration rate on tip resistance

The pile tip resistance-displacement response measured during the load tests performed in tests 2, 3, and 4 are shown in Figures 36, 37, and 38 respectively. The influence of the penetration rate is clearly seen in these figures, although the magnitude of the effect differs between the three tests and depends on the pore fluid used. In tests 2 and 3 where viscous fluid was used, the tip resistance strongly depends on the penetration rate of the pile: the higher the rate, the higher the tip resistance. In test 4 where water was used, the tip resistance in the RLTs is clearly higher than that in the SLT, but the ultimate value of these RLTs is independent of the penetration rate.

Figure 39 shows a generalisation of the dependency of maximum pile tip resistance on the penetration rate of the model pile. The results are taken from all RLTs

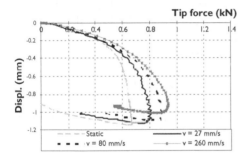

Figure 36 Tip resistance-displacement curves (test 2 low permeability).

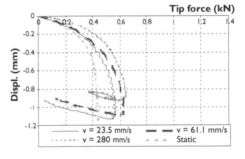

Figure 37 Tip resistance-displacement curves (test 3 low permeability).

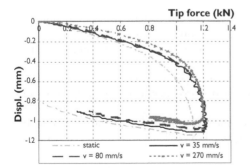

Figure 38 Tip resistance-displacement curves (test 4 high permeability).

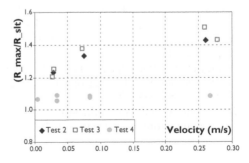

Figure 39 Effect of penetration rate on maximum tip resistance.

performed in tests 2, 3, and 4 with an imposed displacement of 0.1D. In the figure, the maximum tip resistance of a RLT (R_max) is normalised with the value of SLT at the same magnitude of displacement (R_sta). This normalisation eliminates the influence of density in the maximum force. The figure shows that the penetration rate causes an increase in tip resistance of approximately 10% in test 4, whereas the increment varies from 20% to more than 40% in tests 2 and 3, dependent on the rate. The increase of about 10% in test 4 is interpreted as the rate effect and the additional increase in test 2 and test 3 as the influence of pore water pressure. The pile tip resistance is strongly influenced by the penetration rate. The background of these differences will be further discussed in part 2.

7   CONCLUDING REMARKS

The results of the three centrifuge pile load test series have been presented in this chapter. The rapid load tests performed in each centrifuge test are comparable with the prototype rapid load test, both in terms of stress wave number in the pile and pile behaviour. The results presented in this chapter are therefore applicable to the proto-type scale. The following conclusions can be drawn from the test results:

– Due to the high penetration rate of the pile during a RLT, the ultimate pile resist-ance is higher than during a static load test. The effect includes both the load rate effect and the effect of excess pore pressure.
– For shaft resistance, only the effect of pore pressure is observed. At the point of maximum displacement rate effects seem to be absent.
– For tip resistance, both the load rate effect and the effect of excess pore pressure can be seen. The load rate effect is limited (less than 10%). The effect of pore pressure is more significant (more than 30%). The effect of pore pressure depends strongly on the permeability.

REFERENCES

Allard, M.A. (1990). Soil stress field around driven piles. Doctor of philosophy thesis. California Institute of Technology.

Allard, M.A. and Schenkeveld, F.M. (1994). The Delft Geotechnics model pore fluid for centrifuge tests, Proc. of CENTRIFUGE 94, Balkema, Rotterdam, pp. 133–136.

Altaee, A. and Fellenius, B.H. (1994). Physical modeling in sand. Can. Geotech. J. 31(3), pp. 420–431.

Bruno, D. and Randolph, M.F. (1999). Dynamic and static load testing of model piles driven into dense sand. Journal of Geotechnical and Geoenvironmental Engineering, Vol. 125, No. 11, pp. 988–998.

De Nicola, A. and Randolph, M.F. (1994). Development of a miniature pile driving actuator', Proc. Int. Conference Centrifuge '94, Singapore, pp. 473–478.

Eiksund, G. and Nordal, S. 1996. Dynamic model pile testing with pore pressure measurements. Proc. 5 nd Int. Conf. Application of stress-wave theory to piles, Florida, pp. 581–588.

Fuglsang, L.D. and Ovesen, N.K. (1987). The application of theory of modelling to centrifuge studies. Centrifuge in soil mechanics. Craig, W., James, Schofield, A. (eds), Balkema, Rotterdam, pp. 119–138.

Garnier, J., Gaudin, C., Springman, S.M., Culligan, P.J., Goodings, D., Konig, D., Kutter, B., Phillips, R., Randolph, M.F. and Thorel, L. (2007). Catalogue of scaling and similitude questions in geotechnical centrifuge modeling. In International Journal of Physical Modelling in Geotechnics 3.

Gui, M.W. (1995). Centrifuge and numerical modeling of cone penetration tests/piles in sand, PhD Thesis, Cambridge University, UK.

Hölscher, P. (1995). Dynamical response of saturated and dry soils. PhD thesis, Delft University of Technology, Delft, The Netherlands.

Huy, N.Q., Dijkstra, J. and van Tol, A.F. (2005). Influence of loading rate on the bearing capacity of piles in sand. In Gijs Hofmans (Ed.), *16th International Conference on Soil Mechanics and Geotecnical Engineering* (pp. 2125–2128). Rotterdam: Millpress. (TUD)

Huy, N.Q. and van Tol, A.F. and Holscher, P. (2007). A numerical study to the effects of excess pore water pressure in a rapid pile load test. In: V. Soriano, E. Dapena, E. Alonso, J.M. Echave, A. Gens, J.L. de Justo, C. Oteo, J.M. Rodriques-Ortiz, C. Sagaseta, P. Sola and A. Soriano (Eds.), *Proceedings of the 14th European Conference on Soil Mechanics and Geotechnical Engineering* (pp. 247–252). Rotterdam: Millpress. (TUD).

Kusakabe et al. (1998). Research Committee on Rapid Pile Load Test Methods–JGS, Japan. Research activities toward standardization of rapid pile load test methods in Japan. Proc. 2nd Int. Statnamic Loading Test Seminar, Tokyo, Japan, 1998. pp. 219–238.

Mangal, J.K. (1999). Partially–drained loading of shallow foundations on sand. Doctor of philosophy thesis. University of Oxford.

Maeda, Y., Muroi, T., Nakazono, N., Takeuchi, H. and Yamamoto, Y. (1998). Applicability of Unloading-point-method and signal matching analysis on the Statnamic test for cast-in-place pile. Proceedings of the 2nd International Statnamic Seminar, pp. 99–108.

Ovesen, N.K. (1981). Centrifuge tests to determine the uplift capacity of anchor slabs in sand. Proceedings, 10th International Conference on Soil Mechanics and Foundation Engineering, Stockholm, Sweden, Vol. 1, pp. 717–722.

Parkin, A.K. and Lunne, T. (1982). Boundary effects in the laboratory calibration of a cone penetrometer for sand. Proc. of the 2nd. European symp. on Penetration Testing, Amsterdam, Vol. 2, pp. 761–767.

Phillips, R. and Valsangkar, A.J. (1987). An experimental investigation of factors affecting penetration resistance in granular soils in centrifuge modelling. Technical Report No. CUED/D-Soils TR210, Cambridge University, UK.

Sedran, G., Stolle, D.F.E. and Horvath, R.G. (1998). Physical Modelling of Load Tests on Piles. Proceedings of the 2nd International Statnamic Seminar, editors–Kusakabe, O., Kuwabara , F. and Matsumoto, T. 2000, pp. 355–364.

Van der Poel J.T. and Schenkeveld, F.M. (1998). A preparation technique for very homogenous sand models and CPT research. *Centrifuge 98*, Vol. 1, pp. 149–154.

Yamaguchi, Kimura, H. and Fujii, N. (1977). On the Scale Effects of Footings in Dense Sand, Proceeding of the 9th International Conf. on Soil mech. and Foundation Eng, Vol. 1, pp. 795–798.

Chapter 8

Rapid model pile load tests in the geotechnical centrifuge (2): Pore pressure distribution and effects

A.F. van Tol
Delft University of Technology, Delft, The Netherlands
Department of Civil Engineering and Geosciences, Deltares, Delft, The Netherlands

N.Q. Huy
Delft University of Technology, Delft, The Netherlands

P. Hölscher
Deltares, Delft, The Netherlands

SUMMARY

Experimental research on the influence of the loading rate and of generated pore water pressure during rapid load tests is carried out. In a Geotechnical centrifuge, a number of tests on piles in sand were performed. The influence of loading rate and drainage condition are the main item of this research. The testing programme, the test set-up, the test data and results are presented in part 1 of this paper, also in this book. This paper describes the results of the pore pressure measurements and the evaluation of the results. It is concluded that excess pore pressure has a significant effect on the derivation of static capacity from a Rapid Load Test in sand.

I INTRODUCTION

Rapid pile load test methods, such as a Statnamic test or a Pseudo-static test, are considered as an efficient alternative method for static pile load testing because of its fast execution and relatively low costs. During the test, the pile is loaded with a loading duration from 50 ms to 200 ms. For this loading duration, excess pore water pressure is generated in soil closes to the pile even if it is located in sandy soil (Hölscher 1995, Meada 1998, Hölscher, van Tol 2008). The excess pore water pressure may affect either the stiffness or the ultimate bearing capacity of the pile. These effects were studied experimentally by performing a number of rapid pile load tests in a centrifuge. The testing programme, the test set-up, the test data and resulting forces and displacements are presented in part 1 of this paper, (Huy *et al.* 2008). This paper describes the measurement results of the pore pressure in the soil close to the pile toe, the evaluation of the results and a comparison with numerical analyses. Finally, a proposal for a practical approach is presented.

For the convenience of the reader, the instrumentation regarding the measurement of pore pressures is repeated here below. The position of the pore pressure transducers is as follows:

WSM_PL: Pore pressure beneath the pile toe. The pore pressure immediately under the pile toe was measured using a pore pressure transducer integrated into the centre of the pile toe.

WSM 1–4: Pore pressure in the sand bed. The pore pressure was measured at four different positions beneath the pile toe level. The transducers provided information about pore pressure variation as a function of time during a load test. Figure 1 shows the location of the four pore pressure transducers with respect to the pile toe directly after the installation at a depth of 20D (226 mm) below the sand surface. The bottom of the container is 567 mm under the pile toe; the diameter of the container is 600 mm.

The results of three tests are discussed. In test 2 and test 3, the viscous fluid is used as pore fluid. This models sand with a low permeability (silty sand). In test 4, water is used as pore fluid. This models sand with a very high permeability (almost permeability of gravel). For more details, reference is made to Table 4 of Huy *et al.* (2008).

2 RESPONSE OF PORE PRESSURE

The measurements made by the five transducers are presented in this section.

Figures 2(a)(b) to 4(a)(b) present representative pictures of excess pore pressure and the measured values as a function of time, during RLTs in centrifuge test 2 (viscous

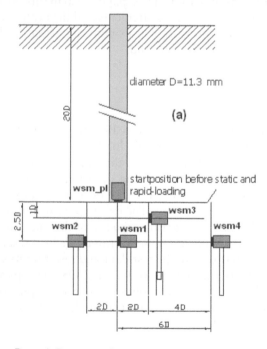

Figure 1 Positions of pore pressure transducers.

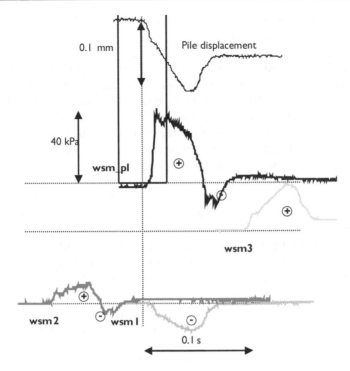

Figure 2a Excess pore pressure during RLT2-1-0.01D; *v* = 2.35 mm/s.

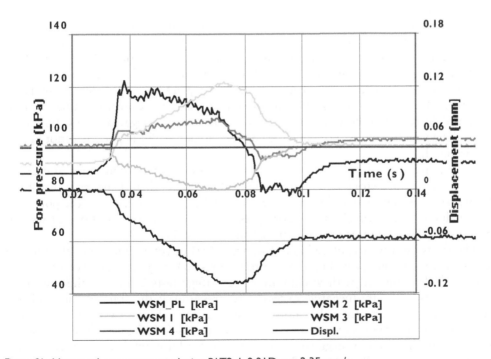

Figure 2b Measured pore pressure during RLT2-1-0.01D; *v* = 2.35 mm/s.

fluid, $I_D = 53\%$). Part (a) of the figure shows the magnitude of pore pressure at the position of the transducer. Part (b) of the figures presents the pore pressure and pile displacement as a function of time in one graph. The pore pressure transducer 4 (wsm4) was defect in test 2 and is therefore not included in the figure.

Figures 2(a) and (b) present results from the slowest and smallest displacement magnitude rapid load test (RLT2-1-0.01D), figures 3(a) and (b) present results from the slow but largest displacement magnitude rapid load test (RLT2-1-0.1D), and figures 4(a) and (b) present results from the fastest and largest displacement rapid load test (RLT2-3-0.1D). The general responses of pore pressure during these RLTs are virtually the same. There is excess pore pressure at all locations. The time needed for that excess pore pressure to attenuate to the static value after the load test (consolidation time) is in the same order of magnitude as the loading duration of the RLT. This implies that the soil is in a partially drained condition during these RLTs. The observed pore pressures in figure 4 are generally higher than those in figure 3. As the penetration rate of RLT2-3-0.1D is higher than the rate of RLT1-1-0.1D, it is concluded that the excess pore pressure depends on the penetration rate.

The tendency of pore pressure response recorded by each pore pressure transducer is described below.

Pore pressure transducers at the pile toe (wsm_pl) and 'diagonally' under the pile toe (wsm 2) show relatively the same response. At the beginning of a RLT, the pore pressure increases before decreasing shortly afterwards. In figures 2(a, b), pore pressure at the pile toe (wsm_pl) slowly decreases with penetration of the model pile but still remains in compression (positive excess pore pressure). When the pile stops moving, the excess pore pressure decreases at a higher rate and becomes lower than the static value (negative excess pore pressure) as the model pile moves upward. After the rapid increase at the start of the test, pore pressure (wsm 2) continues to increase but at a much lower rate. When the pile stops moving, the response is the same as the transducer wsm_pl. This pattern of pore pressure response is only observed in RLTs with an imposed displacement of 0.01*D. For other RLTs with a higher imposed displacement (= 0.02*D), the similarity between wsm_pl and wsm 2 is clearer, as shown in figures 3(a)(b) and 4(a)(b). After the rapid increase to peak positive value, the pore pressure decreases to the value lower than the static value (negative excess pore pressure). The negative excess pore pressure directly underneath the pile toe is not known beforehand.

It is interesting to note that pore pressure at these locations continues to decrease even when the pile stops moving (see figure 3) or even when it moves upward (see figure 4). This implies that there is a region elsewhere in the soil where the pore pressure value is lower, which affects pore pressure response at the locations of wsm_pl and wsm2. This region is probably the shear failure zone.

The transducer wsm1, placed at a distance 2.5*D directly underneath the pile toe, shows a (for us) unexpected pore pressure response. During all RLTs executed in the sand sample saturated with viscous fluid (centrifuge tests 2 and 3), the pore pressure decreases rapidly (negative excess pore pressure) at the start of the load test. At virtually the same time that pore pressure at the pile toe (wsm_pl) reaches its maximum value, the decreasing rate gradually decreases. In figure 2, the excess pore pressure (wsm1) remains negative throughout the load test, while pore pressure starts to increase after a while in figures 3 and 4. This differs from our expectation that positive excess pore pressure would occur at that location.

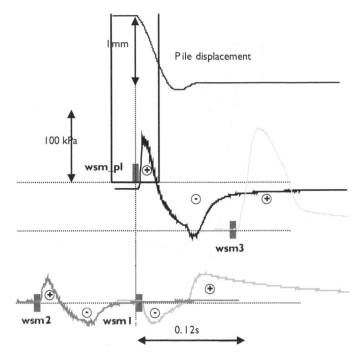

Figure 3a Excess pore pressure during RLT2-1-0.1D; v = 32 mm/s.

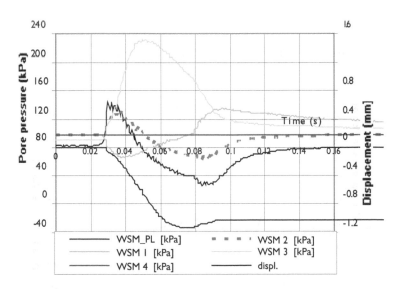

Figure 3b Measured pore pressure during RLT02-1-0.1D; v = 32 mm/s.

The transducer wsm 3 shows consistent pore pressure response at that location during all RLTs, and positive excess pore pressure is observed. However, the increment seems relatively high compared to wsm_pl.

In centrifuge test 3, viscous fluid was also used as the pore fluid. Measurements of pore pressure response show agreement with those in centrifuge test 2. Figures 5(a) (b)

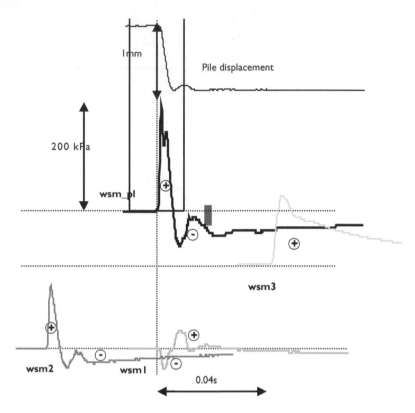

Figure 4a Excess pore pressure during RLT2-3-0.1D; v = 320 mm/s.

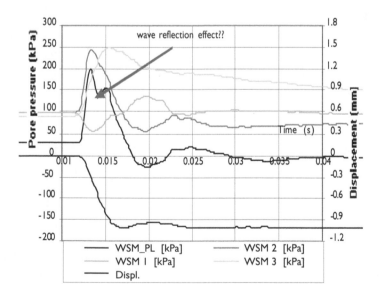

Figure 4b Measured pore pressure during RLT02-3-0.1D; v = 320 mm/s.

presents an example of pore pressure measurements during a RLT in centrifuge test 3 (which has lower initial density than at test 2). The Figure shows the fastest load test with largest displacement amplitude (RLT3-3-0.1D). The responses of pore water pressure transducers are very similar to those seen in figure 4, where the results from a corresponding RLT of the centrifuge test 2 (RLT2-3-0.1D) are shown. It is remarkable, that in test 3 the peak value under the pile toe is higher than in test2, while the values in the transducers 2 and 3 are lower. This will be explained in Section 3 of this paper, using the drainage factor.

Figures 6(a) (b) and 7(a) (b) present the representative results of pore pressure response during RLTs in centrifuge test 4, where water was used as the pore fluid.

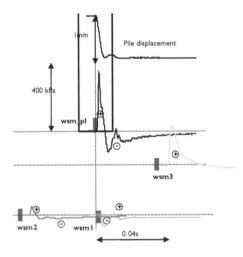

Figure 5a Excess pore pressure during RLT3-3-0.1D; v = 280 mm/s.

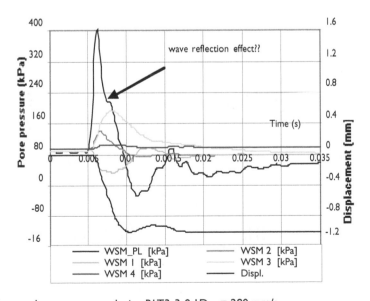

Figure 5b Measured pore pressure during RLT3-3-0.1D, v = 280 mm/s.

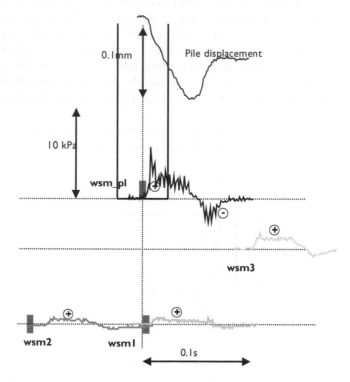

Figure 6a Excess pore pressure during RLT4-1-0.01D; v = 2.35 mm/s.

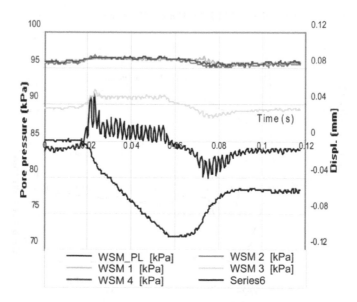

Figure 6b Measured pore pressure during RLT4-1-0.01D; v = 2.35 mm/s.

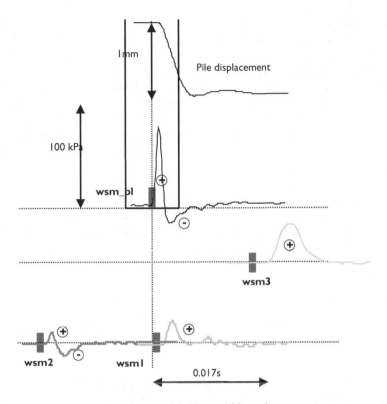

Figure 7a Excess pore pressure during RLT4-3-0.1D, v = 280 mm/s.

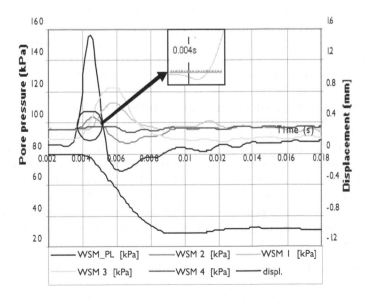

Figure 7b Measured pore pressure during RLT04-3-0.1D, v = 280 mm/s.

Figures 6(a) (b) are the results from the slowest rate and smallest displacement magnitude load test (RLT4-1-0.01D). Figure 7(a) (b) shows the results from the fastest rate and largest displacement magnitude load test (RLT4-3-0.1D). The general responses at the transducer positions wsm_pl, wsm 2, and wsm3 are very similar to those described above. However, the values of excess pore pressure are much smaller and the duration of the excess pore pressure is much shorter. This is understandable from the higher permeability of the sand sample in centrifuge test 4. Only the pore pressure response at the location of the transducer (wsm1) is different. It increases at the start of the load test instead of decreasing immediately as in centrifuge tests 2 and 3, but there is a time lag between the starting pore pressure increment and the start of the load test. The detailed close-up of values from transducer wsm1 included in figure 7b reveals a small decrease in pore pressure at the start of load test RLT4-3-0.1D. It is therefore extremely likely that there is a tendency for negative excess pore pressure at the position of transducer wsm1 during RLTs in centrifuge test 4 that is similar to other measurements in centrifuge tests 2 and 3, but that the value is too small to be seen due to the high permeability of the sand sample.

It should be noted here that excess pore pressure shown in figure 7 attenuates soon after the start of the load test. Nearly halfway through the loading duration of RLT4-3-0.1D, pore pressure values are nearly equal to the static value, i.e. no or minimal excess pore pressure, and soil behaviour can be considered to be in a much more drained condition. This is different from the results in centrifuge tests 2 and 3 because of the viscosity of the pore fluid used in the tests.

3 CHARACTERISTICS OF EXCESS PORE PRESSURE AND THE EFFECT ON TOE RESISTANCE

Overall, the transducers show a consistent pore pressure response during a RLT despite differences in penetration rate and the imposed magnitude of displacement between these RLTs. The pore pressure responses at the locations of the pore pressure transducers during a RLT are qualitative evaluated below. Next, the consequence of the excess pore pressure for the toe resistance due to the penetration rate is examined.

3.1 Characteristics of excess pore pressure

Penetration of a pile into a soil mass will generally cause deformation in the surrounding soil body. If the soil is in a saturated undrained condition, excess pore pressure indicates the tendency of volume change of the soil elements. However, the sand is not in a fully undrained condition during a RLT. During a test, pore fluid flows from the high-pressure region to the lower pressure region in order to equalise the pressure gradient. As a result, excess pore pressure that would exist in an undrained condition will be re-distributed. The magnitude of the excess pore pressure is thus the result of two processes:

1   the generation of pore pressures due to volume changes of the soil:

    a   in zones of compression positive excess pore pressure will be generated;

    b   in zones of predominantly shearing the generated excess pore pressure depends on the density. In the case of the centrifuge test, with a pile installed by pressing in, the soil density around the pile will be high; Shearing will induce negative excess pore pressure.

2   the flow through the saturated soil mass that equalises the pore pressures.

In order to explain the pore pressure response at the location of pore pressure transducers qualitatively, the deformation pattern in the sand region around the pile toe during penetration is considered as reported by Robinksy & Morrison (1964), White (2002) and Nguyen Thanh Chi (2008). White (2002) indicated the formation of a 'nose cone' of high compression and densification of sand particles underneath the pile toe during an increment of pile penetration. When the pile penetrates into the sand, the 'nose cone' moves with the pile and the surrounding sand slides and shears along the edges of the 'nose cone'. Figure 8, taken from Nguyen Thanh Chi (2008) presents the zones of compression and shearing during pile installation, based on this nose cone concept (White 2002) and extended with observation from his PIV-analyses in a 2D test-setup. This pattern helps to explain the observed tendency of the pore pressure responses underneath the pile toe. Based on this pattern, the response of the pore pressure transducer can be described as follows.

The pore pressure plots show that underneath the pile toe (transducer wsm pl) some increase in pore water pressure at the beginning of the loading; but, already after a penetration of 0.1 mm, this changes in a decrease of the pore pressure. The soil directly underneath the pile toe was compressed and pore pressure increased. As the pile moved downwards, shear failure occurred and the sand began to slide along the edges of the 'nose cone' causing negative excess pore pressure in zone 2.

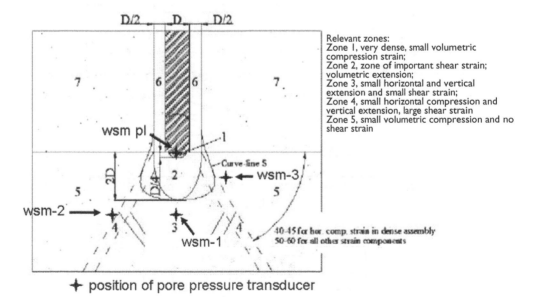

Figure 8 Pore pressure transducers projected in zones of compression and shearing, as observed in a 2D pile installation photo-elastic test (Nguyen Thanh Chi 2008).

Water in zone 1, the 'nose cone' and from the outer regions flows toward the shear zone (zone 2), and pore pressure in the 'nose cone' decreases. The magnitude of the decrease depends on the value of excess pore pressure in the shear zone, which in turn depends on the rate of volume change, and the rate of fluid flow from outside the region to equalise the pressure gradient. Due to the low permeability in tests 2 and 3, the flow (consolidation) continues after the pile stops moving and the source of pore pressure generation has stopped (no further compression and shearing). In test 4 with very high permeability, the flow stops almost directly at the moment of unloading.

Transducer wsm-1 is located in zone 3, close to zone 2. First, a decrease of pore pressure is seen, followed by an increase. The decrease will be due to the shearing in this zone. The later increase might be caused by the excess pore pressure flow from the large compression zone 5. However, this appearance deviates from the measurements from Eiksund & Nordal (1996) as shown in Figure 9.

Transducer wsm-2, located in compression zone 5 close to shearing zone 4, shows first an increase of pore pressure followed by a strong decrease and a negative excess pore pressure, apparently influenced by the shearing zone 4.

Transducer wsm-3 is positioned in zone 5 and shows increase of pore pressure due to the compression in this zone.

3.2   Effects of excess pore pressure

To assess the effect of excess pore pressure, the relation between excess pore pressure underneath the pile toe and toe resistance must be considered. The negative excess

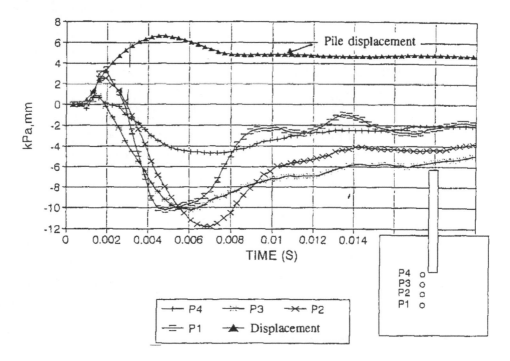

Figure 9 Pore pressure response during a model dynamic pile load test (Eiksund & Nordal 1996).

pore pressure underneath the pile toe during a RLT increases the effective stress in the sand and thus the resistance. However, direct comparison between excess pore pressure values and the increment of total stress under the pile toe does not provide an explanation for the role of excess pore pressure: the value of excess pore pressure is very small in comparison with stress at the pile toe. In RLT2-3-0.1D (Figure 4b) for example, the largest negative excess pore pressure directly underneath the pile toe is less than −80 kPa, whereas the increment in total stress of the RLT2-3-0.1D over the static value of SLT2-3-0.1D is much higher (approximately 2000 kPa, Figure 36 in Huy *et al.* 2008). This is similar to findings by Eiksund & Nordal (1996) and Maeda *et al.* (1998). The increment of 2000 kPa is indeed also caused by the rate effect, but in Huy *et al.* (2008) it is shown that the rate effect plays a minor role in toe resistance. It can therefore be concluded that there is no direct relationship between the value of excess pore pressure at the pile toe and the increment of pile toe resistance due to the penetration rate effect.

However, the classic bearing capacity theory describes the pile base resistance by multiplying the initial effective stress at pile toe level with a bearing capacity factor N_q which is in the order of magnitude of 100 to 200. If this is considered and the measured negative excess pore pressure (−80 kPa) is compared with the initial effective stress at pile toe level (90 kPa) the effect of excess pore pressure is enormous, but this is neither realistic as the excess pore pressure show a large gradient near the pile toe. The role of excess pore pressure should be evaluated by comparing the value of excess pore pressure and the effective stresses at these regions. These stresses are generally not known, i.e. a quantitative comparison is not possible.

3.3   Penetration force and pore pressures

Figure 10 presents an example of the force (right axis) and pore pressure (left axis) at the pile toe as a function of displacement of the pile. The force-displacement relation

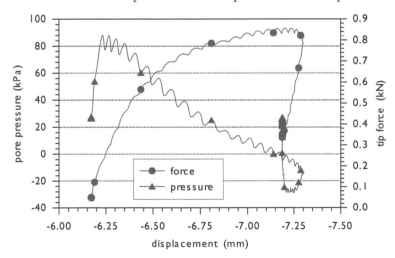

Figure 10 Relation between pore pressure at the pile toe and penetration force as a function of pile displacement, Test 2 penetration velocity is 36 mm/s.

is the load-settlement curve for the pile toe. (Since the force at the pile toe is measured, this requires no inertia correction, which is generally applied when the force at the pile head is measured.) The more negative the displacement in this graph the more the model pile has penetrated. This graph shows that only a limited displacement of around 0.1 mm leads to the maximum pore pressure at the pile toe. Further penetration leads to a gradual decrease in the pore pressure until the maximum penetration is reached. Reversal of the penetration leads momentary to a sharp decrease of the pore pressure, probably due to elastic unloading. After that, the pore pressure increases. The test with a higher penetration velocity (same sample Test 2), gives qualitatively a comparable result, but with some oscillations likely due to dynamic effects, see figure 11.

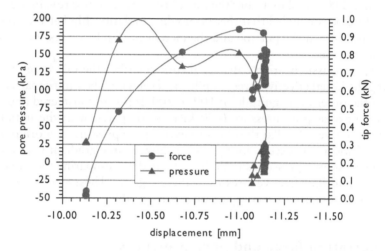

Figure 11 Relation between pore pressure at the pile toe and penetration force as a function of pile displacement, Test 2 penetration velocity is 300 mm/s.

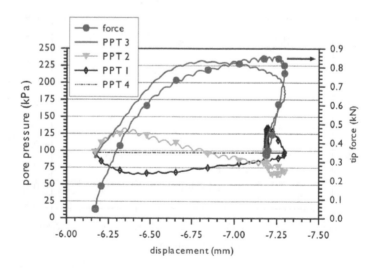

Figure 12 Relation between pore pressure in the sand at various locations and the penetration force as a function of pile displacement, Test 2 penetration velocity is 30 mm/s.

Figure 12 compares the results of the pore pressure transducers in the sand with the toe force as a function of the penetration (Test 2 with penetration velocity of 30 mm/s). It shows that the trend of pore pressure transducer 3 is quite comparable with the penetration force. The other pore pressure gauges have much smaller amplitudes. As mentioned before, wsm-1 results directly in negative pressures and wsm-2 has a trend that is comparable with the pore pressure measured at the pile toe, but only with lower amplitude.

4   IMPLICATIONS FOR THE ANALYSIS OF RAPID PILE LOAD TESTS

4.1   Introduction

This paragraph considers the centrifuge test results in order to assess the implications for the Unloading Point Method (Middendorp 1996, see also Hölscher, van Tol 2008). The discussions focus on the effects on the toe resistance. Toe resistance is predominant in total resistance, not only because important effects are found on toe resistance, but also because of the practical situation of a pile in sand.

The following aspects are discussed:

–   The effects of the penetration rate on pile resistance at the time of maximum displacement, the unloading point resistance (R_up).
–   The effects of the penetration rate on maximum resistance (R_max).
–   The rate dependency law (linear or non-linear).
–   The separation of the rate effect and the pore pressure effect.

The RLTs in the test series were performed with different magnitudes of pile head displacement. Here only the tests with large pile head displacement (0.10D) are considered as they are of importance for the ultimate resistance.

4.2   Rate and pore pressure effects at R_max and at R_up

Figures 13 and 14 respectively plot the dependency of maximum toe resistance (R_max) and the toe resistance value at the time of the unloading point (R_up) on the penetration rate. The values are normalised on the static resistance, in order to remove the influence of the differences in initial density. In Figure 13, data points from centrifuge test 4 show a nearly constant increment of maximum toe resistance between 5% and 10% over the static value, within the tested velocity range of the rapid load test. This increment is the rate effect. By comparing the Figures 13 and 14, it can be concluded that the rate effect increases the maximum toe resistance by approximately 10%, but does not affect the unloading point value, as shown in Figure 14. In the tests with low permeability, the maximum toe resistance and the resistance in the unloading point are much higher. This is due to the pore water pressure effect. The excess pore pressure affects both values.

The trend lines for results from tests 2 and 3 are also plotted in Figures 13 and 14. These data points fit a power law. This means that the rate dependency for pile toe

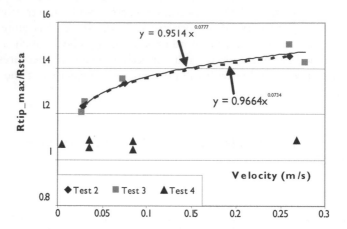

Figure 13 Effect of penetration rate on maximum tip resistance.

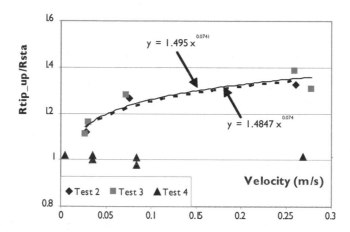

Figure 14 Effect of penetration rate on toe resistance at unloading point.

resistance is non-linear when excess pore pressure plays a role. Since this conclusion is based on a limited number of data points, further verification is necessary.

Figure 15 shows the normalised toe resistance as a function of the dynamic drainage factor (Huy *et al.* 2007, see also Hölscher, van Tol 2008). The dynamic drainage factor depends on the ratio of the duration of the loading and the duration of consolidation. The solid square represents the ratio of maximum pile toe resistance in the RLT over the static value at the same displacement; the open circle represents the ratio of pile toe resistance at the unloading point over the static value at the same displacement. From Figure 15, it is estimated that a drainage factor of approximately 10 can be used to separate the drained side (negligible effect of excess pore pressure) and the partially drained side (effect of excess pore pressure must be considered) for the in-situ rapid load test. However, more test data with a drainage factor between 4 and

100 are required to specify the value. This Figure also shows the practical range of the drainage factor for piles in sand. The following parameters are used to plot these lines: shear modulus $G = 80 \div 160$ MPa, coefficient of permeability $k = 10^{-5} \div 10^{-2}$ m/s, loading duration $T = 80 \div 160$ ms, and pile radius $R = 0.15 \div 0.4$ m.

Finnie and Randolph (1994) studied the effect of the penetration rate in constant rate tests in sand. They concluded that the effect of partial drainage on the penetration resistance can be evaluated against the non-dimensional velocity V, defined as $V = v^* D/c_v$, where v is the penetration velocity, D is the pile diameter, and c_v is the coefficient of consolidation. This non-dimensional velocity is used here, and its value in each RLT is calculated with the penetration velocity v taken as the average velocity of the pile during the loading time of the RLT. Graphs of the normalised resistance against the non-dimensional velocity V are plotted in Figures 16 and 17. It can be seen that an effect of displacement magnitude still exists. Neither the drainage factor nor the non-dimensional velocity incorporates the effect of displacement magnitude.

4.3   Validating the numerical results

Here the results of the centrifuge tests are evaluated against the calculated value of the dynamic drainage factor for each RLT. Figure 18 shows the ratio of maximum pile toe resistance to the static value against the calculated drainage factor. The Figure also includes the result from the numerical studies (Huy *et al.* 2007, see also Hölscher and van Tol 2008). It can be seen that the tendency is very similar, but that the drainage factor where the effect is observed differs. The numerical scheme takes adequately consideration of consolidation during the RLT, hence it shows the same tendency with the experimental results. However, the mechanism for generating excess pore water pressure differs in the numerical scheme with that in the experiment. The effect

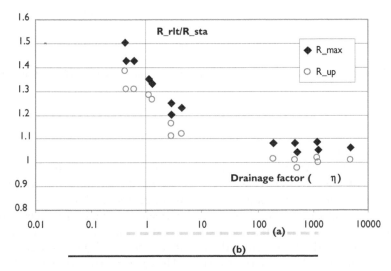

Figure 15 Normalised pile toe resistance against drainage factor.

> *Line (a):* Practical range of drainage factor for Baskarp sand. *Line (b):* practical range of drainage factor for sand.

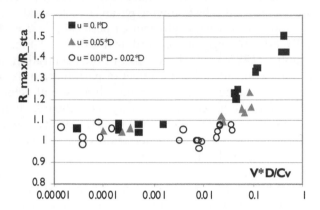

Figure 16 Normalised maximum pile toe resistance against velocity V (*u* = 0.05*D & *u* = 0.1*D).

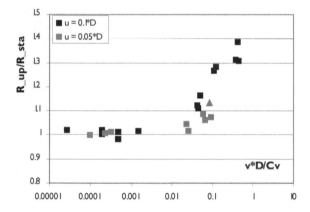

Figure 17 Normalised pile toe resistance at UP time against velocity V (*u* = 0.05*D & *u* = 0.1*D).

of dilatancy is not considered in the numerical scheme, thus the pore pressure only increases as loading increases.

Dilatancy at the shear failure surface may affect the mobilised toe resistance with time. In the centrifuge, this effect will be overestimated in comparison with the prototype situation because of the scaling in the centrifuge: the thickness of the shearbands is not correctly scaled: the shearbands are relatively too thick in the centrifuge (e.g. Stuit 1995). Therefore, the effect leads to a higher value of the boundary for the drainage factor. For the time being, the results of the centrifuge tests are considered as more reliable than the numerical results. However, verification is still required, preferable with field measurements. Other aspects that might influence this effect are the influence of pile installation (verification for piles without soil displacement) and the dilatancy behaviour of the sand.

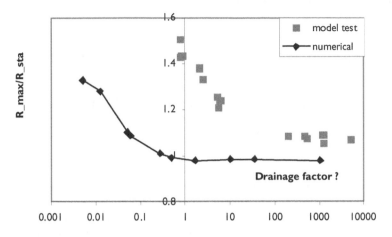

Figure 18 Normalised toe resistance against dynamic drainage factor.

5   CONCLUSIONS

The results of the three centrifuge pile load test series have been presented in this paper. In each test both static and rapid load tests with several rates are carried out alternately. The rapid load tests are comparable with the prototype rapid load test, both in terms of stress wave number in the pile and pile behaviour. The results presented in this paper are therefore applicable to the prototype scale. The test was carried out on soil displacement pile in finely grained sand. The following conclusions can be drawn from the test results:

1   During the penetration of the pile, pore water pressure is generated due to compression and shear failure in the sand. Due to the high penetration rate during a RLT, these cannot dissipate directly during a RLT.

2   The centrifuge tests give insight in the failure pattern of the sand around the pile toe. The empirical observations during pile installation of Nguyen Thanh Chi (2008) are in reasonable agreement with the observations based on the measured pore pressures.

3   The toe resistance of the pile during a RLT is higher than during a static load test. This is due to both the rate effect and the effect of excess pore pressure. The rate effect is limited (less than 10%) and the effect of pore pressure is in the centrifuge tests more significant (maximum of about 30%).

4   The effect of pore pressure can be evaluated using the value of the dynamic drainage factor. The above elaborated results indicate that for engineering practice, knowledge about this drainage factor is vital for analysing an in-situ rapid load test, especially in cases where a large diameter pile is founded in low permeable sand. The shear modulus and permeability should be known for the test site, so that the value of the drainage factor can be estimated.

5   For a soil displacement pile, a dynamic drainage factor value of 10 can be considered as a boundary when deciding whether the effect of excess pore pressure

should be taken into account or not. If the drainage factor is larger than 10, the effect of excess pore pressure is negligible and only the rate effect needs to be considered.

The following conclusions for the application of the unloading point method can be drawn:

If the dynamic drainage factor is larger than 10 (very permeable sand):

6   The UP method assumes that pile resistance at the unloading point is equal to the static value. We observed that the rate effect does not influence the value of pile resistance at the time of maximum pile head displacement (unloading point). Therefore, the UP method can be used in a straightforward way.

7   The maximum resistance is approximately 10% higher then the static value, due to the rate effect. This value is more or less constant within the range of penetration rate. Since this rate is used in practical applications, it could be used as a correction factor for the static value at the time of maximum force.

If the dynamic drainage factor is smaller than 10 (low permeable sand):

8   The excess pore pressure increases both maximum resistance and resistance at the unloading point. Significant errors will therefore arise if the conventional UP method is used without considering this aspect. A correction for the excess pore pressure effect must be applied to predict the static bearing capacity of a pile accurately. However, it is currently not possible to give general recommendations for selecting this correction factor. The experimental curve shown in Figure 14 is the best curve available. This still needs to be validated.

REFERENCES

Eiksund, G. and Nordal, S. (1996). Dynamic model pile testing with pore pressure measurements. Proc. 5th Int. Conf. Application of stress-wave theory to piles, Florida, pp. 581–588.

Finnie, I.M. and Randolph, M.F. (1994). Bearing response of shallow foundations on uncemented calcareous soil, International Conference Centrifuge '94, Singapore, Balkema, Rotterdam, 1: pp. 535–540.

Hölscher, P. (1995). Dynamical response of saturated and dry soils. PhD thesis, Delft University of Technology, Delft, The Netherlands.

Hölscher, P. and van Tol, A.F. (2008) Database of field measurements of SLT and RLT for calibration, Rapid Load Testing on Piles, Hölscher, P. and van Tol, A.F. (eds), Francis Taylor, September 2008.

Hölscher, P., van Tol, A.F. (2008) Introduction, Rapid Load Testing on Piles, Hölscher, P. and van Tol, A.F. (eds), Francis Taylor, September 2008.

Hölscher, P., van Tol, A.F., Huy, N.Q. (2008) Influence of rate effect and pore water pressure during Rapid Load Test of piles in sand, Rapid Load Testing on Piles, Hölscher, P. and van Tol, A.F. (eds), Francis Taylor, September 2008.

Huy, N.Q., van Tol, A.F. and Hölscher, P. (2008) Rapid Model pile load tests in the geotechnical centrifuge, part 1, Rapid Load Testing on Piles, Hölscher, P. and van Tol, A.F. (eds), Francis Taylor, September 2008.

Huy, N.Q., Hölscher, P. and van Tol A. F. (2007) A numerical study to the effects of excess pore water pressure in a rapid pile load test, Proc. XIV European Conference on Soil Mechanics and Geotechnical Engineering, Madrid, Spain, 24–27 September 2007.

Meada, Y., Muroi, T., Nakazono, N., Takeuchi, H. and Yamamoto, Y. (1998) Applicability of Unloading-point-method and signal matching analysis on the Statnamic test for cast-in-place pile. Proceedings of the 2nd International Statnamic Seminar, pp. 99–108.

Middendorp, P., Bermingham, P. and Kuiper, B. (1992) STATNAMIC Load Testing of Foundation Piles. Proc. Fourth International Conference on the Application of Stress Waves on Piles, the Hague, Balkema, pp. 581–588.

Nguyen Thanh Chi (2008). Investigation into soil displacement near a jacked-in pile in sand. Master Thesis, Section Geo-Engineering, Delft University of Technology, Delft, Netherlands.

Robinksy, E.I. and Morrison, C.F. (1964) Sand displacement and compaction around model friction piles. Canadian Geotechnical Journal, 1(2): 81–93, 1964.

Stuit, H.G. (1995) Sand in the geotechnical centrifuge, PhD-thesis Delft University of Technology, 1995.

White, D. (2002) An investigation into the behaviour of pressed in piles. PhD Thesis, University of Cambridge, UK.

Chapter 9

Database of field measurements of SLT and RLT for calibration

P. Hölscher
Deltares, Delft, The Netherlands

A.F. van Tol
Delft University of Technology, Delft, The Netherlands
Department of Civil Engineering and Geosciences, Deltares, Delft, The Netherlands

SUMMARY

The usage of the rapid load test (RLT) as a replacement of a static load test (SLT) might have economic benefits. However, it should be proven that this alternative is justified.

Since a RLT requires a more complex interpretation then the SLT, the interpretation rules for the RLT must be evaluated. This paper focuses on the available empirical data for the validation of interpretation methods. Data from literature are presented. This reveals that the interpretation for tests in clay is indeed problematic. For piles in sand the installation method seems to play a role.

Additional field tests at Waddinxveen are presented. Special reference is made to the occurrence of pore water pressure around the pile toe in sand. It is concluded that this effect plays a role in a RLT and might hinder the interpretation of a test, surely in finer grained materials.

1 INTRODUCTION

The usage of the rapid load test (RLT) as a full replacement of a static load test (SLT) might have economic benefits. However, it should be proven that the RLT is a reliable test for the determination of pile behavior. Empirical data are essential to proof the applicability of RLT.

A RLT requires more interpretation than a SLT. Dynamic effects, rate effects and pore water pressure may play a role in the interpretation. Moreover, the type of soil, the type of pile and installation method will play a role. All these aspects must be examined carefully. To do so, a relatively extended set of empirical data is required.

The execution of tests that compare the SLT and RLT are expensive, and therefore scarce. Available results must be treated with care.

This paper presents an overview of available results where SLT and RLT are compared. It discusses the importance of careful evaluation of available data. Finally, it presents the need to establish an international database with measured data, which can be used to validate interpretation models.

2   GENERAL REVIEW OF THE RESEARCH

The research started with collecting available data on field measurements in literature. The database of McVay *et al.* 2003, is a good starting point. This database is extended with more recent data, and some results which where not (yet) published in literature.

During this extension, it turned out that the tested piles not always had geotechnical failure during testing. Two groups are formed: tests with and without failure. If only the stiffness is of interest, a test without failure is good enough.

In general, such databases are based on the interpretation method used by the author of the reference. This is an important disadvantage. The first reason is the fact that the interpretation method is not always clearly defined, and might be subject to changes during time. The second reason is that other, more recently developed methods cannot be evaluated by such results, since the originally measured data is not available. A database with original measurements has not such disadvantages and is preferred. A description of requirements is made to start the development of an extended database.

3   CREATION OF DATABASE WITH BEARING CAPACITIES

3.1   Data from literature

The database, originally developed by McVay *et al.* (2003), is extended with more recent case studies from literature. Table 3 (in the annex of this paper) presents the data of McVay in full extend, while Table 1 presents a summary of the results of McVay *et al.* For the calculation of the static capacity from the RLT, the Unloading Point Method (UPM, Middendorp 1992) was used. Table 1 presents the McVay data as the result of the SLT divided by the result of the RLT. This means that the result of a RLT in average must be multiplied with this empirical factor to obtain the result of the SLT. Here, this empirical factor is called R, so

$$R = \frac{F_{slt}}{F_{rlt}} \Leftrightarrow F_{slt} = RF_{rlt} \tag{1}$$

Table 1 gives also the standard deviation and the number of data in the table. This last number is required to calculate the reliability intervals with the Student-t distribution.

Table 1 Summary empirical data of McVay (2003).

Material	Clay	Sand	Silt
Empirical factor	0.53	0.92	0.70
Standard deviation	0.49	0.17	0.19
Coefficient of Variation	0.92	0.18	0.27
Number of cases	6	9	14

3.2   Extension of these data

In order to update the results of McVay *et al.* results were searched of more recent tests. Both open literature as reports are used for this extension. It turned out that in this data not always failure had been reached during the tests. This must be considered, since the rate effect and the damping effect differs during initial loading and failure.

Table 4 (at the annex of this paper) gives an overview of the cases where real failure (displacement above 5% of pile diameter) occurred during the Rapid Load Test. Generally, the extension confirms for piles in sand and clay the empirical knowledge that R in sand is about 0.90 and in clay 0.50.

Inspection of the results in Table 4 shows that most of the cases are obtained at three sites (Maasvlakte in the Netherlands, Limelette in Belgium and Sint-Katelijne-Waver in Belgium). The tests at the Maasvlakte (Opstal *et al.* 1996) give results for two types of prefabricated piles in predominantly sand, a empirical factor R of 0.84– 0.96 was obtained. This agrees with the results of McVay *et al.* The tests at Limelette (Holeyman *et al.* 2003) show results for three types of cast in-place piles in sand. The scatter is relatively high. This suggests that for cast in-place piles care should be taken for the choice of the model parameter R. In Sint-Katelijne-Waver (Holeyman *et al.* 2001), several types of cast in-place are tested in clay. In this case, the scatter is much lower.

Table 5 (in the annex of the paper) gives an overview of the cases where no real failure (displacement lower than 5% of pile diameter) occurred during the Rapid Load Test. As these piles were not tested up to failure the results, in term of a comparison of the capacities in SLT and RLT, does not make sense. It might only give some information about the ratio of pile stiffness in a SLT and a RLT. The empirical factor on stiffness R_{sf} for these tests in sand is about 0.95 on average and in clay about 0.80. These values are higher then at ultimate capacity.

4   ADDITIONAL DATA FROM WADDINXVEEN

4.1   Evaluation of results per pile type

The building of the IFCO-office in Waddinxveen has a foundation with several pile types. IFCO, a geotechnical consultant firm, monitors the long-term behavior of foundation piles. In the mid-nineties, these piles were tested rapidly, using the PSPLT equipment of Fundex. This had been done shortly after installation. After completion of the building (about 1 year), the piles were tested statically, using the weight of the building as static mass. These static tests were carried out much later; the piles already carried the load of the building.

About 28 piles were tested by the PSPLT (Middendorp 2008) (Hölscher *et al.* 2008b). The analysis of these old data was hindered by the unloading of the PSPLT frame. When the equipment releases the mass, an unloading wave is generated at the surface. For large falling heights, this unloading wave reached the displacement measurement equipment before the mass loads the pile. Such measurements must be eliminated. This quandary reduced the number of useful tests substantially.

Table 6 shows a summary of the results in the format from Table 4. Two types of piles are analyzed: precast concrete and CFA. The piles have their toe in a bearing layer of sand. The Interpretation is carried out according to the hyperbolic description presented by Middendorp *et al.* (2008). In the column "Rapid test", no model factor R is included.

The following observations are made:

– the scatter for prefab concrete pile is lower than for CFA piles;
– the modal factor R is for the prefab piles order 0.9, which confirms the data from Table 1 for piles with the toe in sand;
– For the CFA piles, the factor of about 1.1 is much higher than expected.

This last observation suggests that the piles that are installed with soil removal (CFA and bored piles), have a deviant behavior during a RLT, compared with piles with soil displacement. This might result in a different model factor R.

4.2 Pore water pressures under instrumented piles

The research described in the paper of (Hölscher *et al.* 2008a) shows the importance of the pore water pressure for piles in sand. Only a few field measurements with pore water pressure during rapid load testing are known from literature.

– (Hölscher & Barends 1992) showed that during a Statnamic test on a pile in sand, pore water over- and under pressures are generated at about 70 cm from the pile toe. The maximum value is order 100 kPa. For the pile (with an equivalent diameter of 28 cm) the consolidation time is about 150 ms. From their measurements, it is concluded that in this case the soil is expected to behaves partly drained.
– (Maeda, Muroi *et al.* 1998) measured the pore water pressure near to the toe of a cast-in-place pile with 1.2 m diameter. The pile (length 13.4 m, diameter 1.2 m) was installed with its' toe 1.2 m in alluvial gravel sand layer. At 0.5 m and 1.1 m from th pile toe, they measured values up to 80 kPa, with consolidation time of order 400 ms, which is longer than the duration of the test. This test can be considered as almost undrained.
– (Hajduk, Paikowsky *et al.* 1998) measured the pore water pressure in a sand layer near the pile toe. The pore water pressure was extremely small and consolidation took about 10–15 minutes. This is extremely long for a sand layer, but no explanation is given.
– (Matsumoto 1998) measured the pore water pressure around an open-ended pile driven into soft rock. Along the pile shaft, negative values are observed, below the pile toe positive values are observed. The peak values are 200 kPa at 1 m from the pile toe, the consolidation time is about 200 ms.
– At Waddingsveen (The Netherlands) we carried out a full-scale field test consisting of two precast concrete piles (350×350 mm²) driven 2 m into the sand. The pore pressure was measured at the level of the pile toe at a distance of 0.60 m during a RLT test. The measured excess pore water pressure in this transducer was about 130 kPa, which is in good agreement with the results of Hölscher and Barends 1992.

From these measurements and the centrifuge test, reported by van Tol *et al.* (2008), it is concluded that pore water pressures are generated during a rapid load test in sand. This excess pore pressures will have two effects:

- an additional pore water force against the pile toe.
- a positive or negative influence on the actual strength of the sand around the pile toe depending on the sign (positive or negative) of the excess pore pressure.

In the centrifuge tests negative excess pore pressure was observed around the pile toe increasing the strength of the soil.

For the further interpretation of the test data two observations are made:

- The negative excess pore pressure is the result of dilatancy and might be over-estimated in the centrifuge in relation to prototype situation, as dilatancy scales different then the other parameters in centrifuge testing
- The centrifuge tests were carried out on displacement piles, while the pore pressures next to bored and CFA piles might show a deviating behaviour because of loosening of sand close to these piles during installation.

5   RE-ORDERING THE AVAILABLE DATA

All available data are re-ordered. New tables are created for the following cases:

- piles with soil displacement in clay (Table 7)
- piles with soil displacement in sand (Table 8)
- piles with soil removal in sand (Table 9)

Only cases with geotechnical failure are included in these tables. For the case "piles with soil removal in clay" no empirical data are available. In clay, we expect a minor influence of the installation.

With respect to the results in Table 9, it is noted that this number refers only to three sites; at the Waddinxveen site, six piles are tested. This means that the table presents only three independent cases. In the other table, similar dependency of results is observed. Analysis of Table 7, Table 8 and Table 9 shows:

- The piles in clay (Table 7) are influenced by the large number of tests in Belgium at Sint-Katelijne-Waver, which has a rather low variation. In this case, the influence of the variance at one site seems to be small.
- The soil displacement piles in sand (Table 8) are not strongly influenced by the measurements at the Maasvlakte in the Netherlands, since these measurements have a mean value and variation that is comparable with the other results. This suggests that the variation between sites is rather small.
- The piles in sand with soil removal (Table 9) show that the variance is strongly influenced by the lower variance found at the Waddinxveen site. This suggests that the differences between the sites are relatively important.

Table 2 presents an overview of the empirical factors, which were derived from the results. In the last line, the number of sites is given. This is the number that should be used for the degrees of freedom the Student-t test for reliability. For clay, the high coefficient of variation shows that the application of RLT in clay should be handled with care.

Table 2 Summary of all data in literature.

Pile type material	Displacement clay	Displacement sand	Bored and CFA sand
Data in	Table 7	Table 8	Table 9
Empirical factor	0.66	0.94	1.11
Standard deviation	0.32	0.15	0.46
Coefficient of Variation	0.49	0.16	0.42
Number of cases	12	19	8
Number of sites	6	9	3

6 EXTENDED DATABASE FOR RELIABILITY

The data from literature cannot be used to compare different interpretation methods, such as the method from Sheffield and the unloading point method. For this comparison (original) measurement data is required.

Such a general extended database with independent cases will be constructed. This database must contain a description of the site, the piles, the installation and the tests carried out, together with the originally measured data. The field test in Waddinxveen (Brassinga and, Middendorp, this book, Middendorp 2008) are examples which will be included in the database. Other field measurements, e.g. the field tests in Belgium at Limelette (Holeyman, Charue 2003) and Sint-Katelijne-Waver (Holeyman, Couvreur, Charue, 2001) are excellent tests for this database. This database offers the possibility to judge the reliability of newly developed interpretation methods.

6.1 Description of database

The format for the cases in the database is clearly defined and the minimum requirements are given as well. Both the format and the accepted cases are published on the internet (Hölscher 2007). All measurement data must be available digitally, according to the definitions in the Standard. The requirements are less strict than those defined in the Standard, in order to make it possible to add older measurements and measurements with some minor shortcoming into the database.

6.2 Requirements

The minimum information required to accept a measurement for the database:

1   A site description with layer thickness and layer type and a near by measurement (e.g. SPT or CPT)
2   A description of the pile(s): type, length, cross section and dimensions, material, stiffness
    If the static test and the rapid test are not on the same pile, a remark why the piles are comparable. If the piles are not identical, a list with differences between the piles
3   A description of the installation: method, date of installation, toe level after installation

4   A description of the static test: date of execution, location of the pile, reaction method, working order, loading history of the pile
    It is essential that the duration between load steps is long enough to allow full creeping
5   A description of the Rapid Load Test: date of execution, apparatus used, working order, number of cycles, delay between cycles, location of the pile, loading history of the pile
6   The measurement data for the static test (in ascii)
7   The measurement data for the rapid test (force, displacement, acceleration) as a function of time with proper sampling frequency (in ascii).

The data will be reworked to the following documents:

–   a report with information (points 1 to 5)
–   a gef-file for the static test on each pile
–   a gef file for each loading cycle of the rapid test on each pile (note: gef = geotechnical exchange file see www.deltares.nl)

The report contains graphs of the measured data, including a load-displacement diagram for each load cycle of the rapid load test based on the measured data (without) any correction. The publication of the data is free of rights. If the static and rapid tests are carried out and reported according the requirements of the Euro codes, these requirements are met automatically. In that case, these two reports are published in full content.

7   CONCLUSION

In order to improve the applicability of the rapid load test on piles empirical data of field tests are essential for interpretation of the test results. A basic database with final results on bearing capacity is extended. Relatively high variances are found. The situation for piles in clay seems more complicated (and less reliable) than for piles in sand. The type of pile installation (with or without soil displacement) plays a role.
    The field tests in Waddinxveen are elaborated. The pore water pressure measurements at the same location shows the generation of pore water pressure. This aspect must be taken into account.
    An extended database for judgment of interpretation methods is suggested. On the long term, these activities lead to an objective judgment of the usage of the method as a valuable and economical replacement of the static load test.

REFERENCES

Briaud, J.L., Ballouz, M. and Nasr, G. (2000). 'Static capacity predictions by dynamic methods for three bored piles', Journal of Geotechnical and Geoenvironmental engineering, July 2000, pp. 640–649.
Brown, M.J. (2004). 'The rapid load testing of pile in fine grained soils', PhD-thesis at the University of Sheffield, Sheffield, UK, March, 2004.

Ealy, C.D. and Justason, M.D. (1998). 'Statnamic and static load testing of a model pilegroup is sand', Kusakabe, Kuwabara & Matsumoto (eds); Statnamic loading test '98; Balkema, Rotterdam, the Netherlands, 2000.

Hajduk, E.L., Paikowsky, S.G. *et al.* (1998). 'The behavior of piles in clay during Statnamic', dynamic and different static load testing procedures', 2nd int. Statnamic seminar. Tokyo, Japan.

Heijnen, W.J. and Joustra, K. eds (1996). Application of stress wave theory to piles: Test results, Balkema, Rotterdam, the Netherlands.

Holeyman, A. and Charue, N. (2003). International pile capacity prediction event at Limelette, Belgian Screw pile Technology, design and recent developments, Maertens & Huybrechts (eds), Balkema, Lisse, NL, May 2003.

Holeyman, A., Couvreur, J.M. and Charue, N. (2001). Results of dynamic and kinetic pile load tests and outcome of an International prediction event; Screw piles, Technology, installation and design in stiff clay; Holeyman, A. (ed), Balkema, Lisse, NL, March 2001.

Hölscher, P. and Barends, F.B.J. (1992). Statnamic load testing of foundation piles. Fourth Intern. Conf. on application of stress wave theory to piles. The Haue, the Netherlands, Balkema.

Hölscher, P., van Tol, A.F. and Huy, N.Q. (2008a). Influence of rate effect and pore water pressure during Rapid Load Test of piles in sand, Recent developments in rapid load testing on piles, Hölscher, P., van Tol, A.F. (eds), Francis Taylor, September 2008.

Hölscher, P., van Tol, A.F. and Middendorp, P. (2008b) European standard and guideline for Rapid Load Test, proc. 8th Int. Conf. on the application of stress wave theory in piles, Lisbon, Portugal, September 2008, to be published.

Horvath, J.S., Trochalides, T., Burns, A. and Merjan, S. (2004). 'Axial compressive capacities of a new tapered steel pipe pile at the John F. Kennedy international airport', proc. 5th Int. Conf. on case histories in geotechnical engineering; New York, NY, April 13–17.

Maeda, Y., Muroi, T. *et al.* (1998). Applicability of unloading-point-method and signal matching analysis on Statnamic test for cast-in-place pile. 2nd Int. Statnamic seminar. Tokyo, Japan.

Matsumoto, T. (1998). Finite element analysis of Statnamic loading test of pile. 2nd Int. Statnamic seminar, Tokyo, Japan.

McVay, M., Kuo, C.L. and Guisinger, A.L. (2003). Calibrating resistance factors for load and resistance factor design for statnamic load testing, report University of Florida, 491 045 048 2312, March 2003.

Middendorp, P., Bermingham, P. and Kuiper, B. (1992). Statnamic load testing of foundation piles, "Proc. 4th Int. Conf. Appl. Stress-Wave Theory to Piles, The Hague, Sept. 1992", Rotterdam, Balkema, 1992, pp. 581–588.

Middendorp, P. (2008). Report Rapid Load Testing Waddinxveen, Profound, R07DM010. PM.RLT, draft May 2008.

Middendorp, P., Beck, C. and Lambo, A. (2008). Verification of statnamic load testing with static load testing in a cohesive soil type in Germany, proc. 8th Int. Conf. on the application of stress wave theory in piles, Lisbon, Portugal, September 2008, to be published.

Opstal, A. Th. P.J. and van Dalen, J.H. (1996). Proefbelasting op de Maasvlakte (in Dutch), Cement, no. 2, pp. 53–62 (in Dutch).

Presten, M.R. (2001). Report on Pier testing program/Fundex PLT with appendices, October 26, 2001.

Schmuker, C. (2005). 'Comparison of static load tests and statnamic load tests' (in German: Vergleich statischer und statnamischer Pfahlprobebelastungen), MSc-Thesis Biberach University, Germany.

Van Tol, A.F., Huy, N.Q., Hölscher, P. (2008). Rapid model pile load tests in the geotechnical centrifuge (2): Pore pressure distribution and effect, Rapid Load Testing on Piles, Hölscher, P., van Tol, A.F. (eds), Francis Taylor, September 2008.

APPENDICES

Table 3 Overview McVay database (2003).

Location	Pile type	Soil type	Test order	Static test kN	RLT test kN
Contraband T114, USA	Driven PC	clay	RLT-STL	1830	3070
Nia TP 5 & 6B, USA	Pipe	clay	STL-RLT	2190	2600
Amherst 2, USA	Driven steel	clay	next	1214	1244
Amherst 4, USA	Driven steel	clay	next	965	1617
S9102 T2, CAN	Pipe	clay	next	1040	2550
S9306 T2, USA	Pipe	clay	next	1360	892
BC pier 5, USA	Driven PC	sand	RLT-STL	3500	3957
BC pier 10, USA	Driven PC	sand	STL-RLT	3380	5000
BC pier 15, USA	Driven PC	sand	STL-RLT	3820	3322
Shonan T5, JPN	Driven bored	sand	STL-RLT	446	489
Shonan T6, JPN	Driven pipe	sand	RLT-STL	1100	1042
S9004 T1, CAN	AC	sand	next	1310	1350
S9209 T1, USA	Driven steel	sand	STL-RLT	7130	6370
YKN-5, JPN	Driven PC	sand	STL-RLT	2770	2700
Hasaki-6, JPN	Pipe	sand	STL-RLT	1890	1490
Noto, JPN	Steel pipe	soft rock	STL-RLT	4380	5087
ashaft10	DS	silt	STL-RLT	1420	2530
ashaft8	DS	silt	STL-RLT	1700	1680
ashaft7	DS	silt	STL-RLT	2230	2430
ashaft5	DS	silt	STL-RLT	2800	2230
ashaft3	DS	silt	STL-RLT	1013	1200
ashaft2	DS	silt	STL-RLT	2230	2030
ashaft1	DS	silt	STL-RLT	2400	2050
NIA TP 12a	pipe	silt	next	1230	1285
NIA TP 12b	pipe	silt	next	1300	950
NIA TP 13a	pipe	silt	next	1210	1225
NIA TP 13b	pipe	silt	next	1300	1136
NIA TP 910a	pipe	silt	next	1810	1900
NIA TP 910b	pipe	silt	next	2380	1890
S9010T1	DP	silt	STL-RLT	2470	2360

Notes: DS = Drilled Shaft, DP = Driven Pile.

Table 4 Results from literature with failure.

Source	Location	Pile type	Soil type	Test order	Static test kN	Rapid test kN
Ealy, Justason 1998	laboratory scale	steel pipe	wet sand	unknown	240	250
Briaud 2000	sand, pile 2	bored pile with defects	sand	Statnamic, (dynamic,) static	1602	2460
Presten 2001	mixed, pile site 1 middle	drilled pile	sand/clay layers	pseudo-static, static	5200	3300
Heijnen, Joustra 1996	Delft, NL, pile 5	concrete Driven	clay with sand layers, tip in sand	static, quasi-static	1200	1800
Opstal e.a. 1996	Maasvlakte NL, pile 1	steel HP heavy point, open end	sand with clay layers	unknown	5350	5800
Opstal e.a. 1996	Maasvlakte NL, pile 2	steel HP heavy point, closed end	sand with clay layers	unknown	6500	7100
Opstal e.a. 1996	Maasvlakte NL, pile 3	steel HP heavy point, closed end	sand with clay layers	unknown	6400	7500
Opstal e.a. 1996	Maasvlakte NL, pile 6	concrete with casing	sand with clay layers	unknown	4400	5000
Opstal e.a. 1996	Maasvlakte NL, pile 8	concrete with casing	sand with clay layers	unknown	4630	4800
Opstal e.a. 1996	Maasvlakte NL, pile 10	concrete with casing	sand with clay layers	unknown	4290	5100
Holeyman e.a. 2003	Limelette B	de Waal	top clay, deep sand	nearby pile	2200	2850
Holeyman e.a. 2003	Limelette B	Fundex	top clay, deep sand	nearby pile	2850	3000
Holeyman e.a. 2003	Limelette B	Omega	top clay, deep sand	nearby pile	2700	2550
Holeyman e.a. 2001	Sint-Katelijne-Waver B	Prefab	clay	nearby pile	1364	3110
Holeyman e.a. 2001	Sint-Katelijne-Waver B	Fundex	clay	nearby pile	1216	3033
Holeyman e.a. 2001	Sint-Katelijne-Waver B	de Waal	clay	nearby pile	1258	2580
Holeyman e.a. 2001	Sint-Katelijne-Waver B	Olivier	clay	nearby pile	1722	3100
Holeyman e.a. 2001	Sint-Katelijne-Waver B	Omega	clay	nearby pile	1263	2454
Holeyman e.a. 2001	Sint-Katelijne-Waver B	Atlas	clay	nearby pile	1637	2766

Note: The column "Rapid test" shows the results of the method without the model factor *R*.

Table 5 Test results for cases without failure.

Source	Location	Pile type	Soil type	Test order	Static test kN	Rapid test kN
Horvath e.a. 2004	New York USA	tapered steel with concrete	sand with clay layer	unknown	850	800
Briaud 2000	NGES-TAMU, pile 4	bored pile	sand	Statnamic, (dynamic,) static	4004	4490
Briaud 2000	NGES-TAMU, pile 7	bored pile	clay	Statnamic, (dynamic,) static	3025	3150
Brown, 2004	Grimsby, in situ pile	auger bored pile	clay	pseudo-static, static	2000	2346
Hajduk 1998	Newburry USA, TP 2	driven steel pipe (open?)	clay	static, pseudo-static	800	1200
Hajduk 1998	Newburry USA, TP 3	driven concrete	clay	static, pseudo-static	1025	1450
Opstal e.a. 1996	Maasvlakte NL, pile 1	steel HP heavy point, open end	sand with clay layers	unknown	3250	3800
Opstal, e.a. 1996	Maasvlakte NL, pile 2	steel HP heavy point, closed end	sand with clay layers	unknown	3200	3700
Opstal, e.a. 1996	Maasvlakte NL, pile 3	steel HP heavy point, closed end	sand with clay layers	unknown	2700	3050
Opstal, e.a. 1996	Maasvlakte NL, pile 6	concrete with casing	sand with clay layers	unknown	2900	3300
Opstal, e.a. 1996	Maasvlakte NL, pile 8	concrete with casing	sand with clay layers	unknown	3200	2900
Opstal, e.a. 1996	Maasvlakte NL, pile 10	concrete with casing	sand with clay layers	unknown	3050	2800
Holeyman e.a. 2003	Limelette B	Prefab	top clay, deep sand	nearby pile	2800	3400
Holeyman e.a. 2003	Limelette B	Atlas	top clay, deep sand	nearby pile	2850	3400
Holeyman e.a. 2003	Limelette B	Olivier	top clay, deep sand	nearby pile	2000	1750
Schmuker	P2, Minden BRD	Jacbo SOB	clay	Statnamic, static	3400	4100
Schmuker	P12, Minden BRD	Jacbo SOB	clay	Statnamic, static	3600	4500

Table 6 Results of analysis Waddinxveen tests.

Pile number	Pile type	Relative displacement	Static test kN	Rapid test kN
pile 4	prefab concrete	9%	640	656
pile 6	prefab concrete	30%	700	731
pile 7	prefab concrete	14%	920	1074
pile 9	prefab concrete	7%	660	741
pile 48	CFA	10%	920	742
pile 49	CFA	11%	840	1221
pile 50	CFA	12%	820	649
pile 51	CFA	17%	540	503
pile 52	CFA	19%	840	754
pile 53	CFA	9%	840	653

Notes:
All data from Middendorp 2008.
All piles in top-layer of clay with toe in sand.
All piles are loaded rapidly (after installation) and statically after construction of the complete building.
The column "Relative displacement" is the displacement at the unloading point relative to the pile diameter.
The column "Rapid test" shows the results of the method without the model factor R.

Table 7 Overview results displacement piles in Clay.

Source	Location	Pile type	Test order	Static test kN	Rapid test kN	Model factor
McVay et al. 2003	Contraband T114, USA	Driven PC	RLT-SLT	1830	3070	0.60
McVay et al. 2003	Nia TP 5&6B, USA	Pipe	SLT-RLT	2190	2600	0.84
McVay et al. 2003	Amherst 2, USA	Driven steel	next	1214	1244	0.98
McVay et al. 2003	Amherst 4, USA	Driven steel	next	965	1617	0.60
McVay et al. 2003	S9102 T2, CAN	Pipe	next	1040	2550	0.41
McVay et al. 2003	S9306 T2, USA	Pipe	next	1360	892	1.52
Holeyman e.a. 2001	Sint-Katelijne-Waver B	Prefab	nearby pile	1364	3110	0.44
Holeyman e.a. 2001	Sint-Katelijne-Waver B	Fundex	nearby pile	1216	3033	0.40
Holeyman e.a. 2001	Sint-Katelijne-Waver B	de Waal	nearby pile	1258	2580	0.49
Holeyman e.a. 2001	Sint-Katelijne-Waver B	Olivier	nearby pile	1722	3100	0.56
Holeyman e.a. 2001	Sint-Katelijne-Waver B	Omega	nearby pile	1263	2454	0.51
Holeyman e.a. 2001	Sint-Katelijne-Waver B	Atlas	nearby pile	1637	2766	0.59

Table 8 Overview results displacement piles in sand.

Source	Location	Pile type	Test order	Static test kN	Rapid test kN	Model factor
McVay et al. 2003	BC pier 5, USA	Driven PC	RLT-SLT	3500	3957	0.88
McVay et al. 2003	BC pier 10, USA	Driven PC	SLT-RLT	3380	5000	0.68
McVay et al. 2003	BC pier 15, USA	Driven PC	SLT-RLT	3820	3322	1.15
McVay et al. 2003	Shonan T5, JPN	Driven bored	SLT-RLT	446	489	0.91
McVay et al. 2003	Shonan T6, JPN	Driven pipe	RLT-SLT	1100	1042	1.06
McVay et al. 2003	S9004 T1, CAN	AC	next	1310	1350	0.97
McVay et al. 2003	S9209 T1, USA	Driven steel	SLT-RLT	7130	6370	1.12
McVay et al. 2003	YKN—5, JPN	Driven PC	SLT-RLT	2770	2700	1.03
McVay et al. 2003	Hasaki—6, JPN	Pipe	SLT-RLT	1890	1490	1.27
Heijnen, Joustra, 1996	Delft, NL, pile 5	concrete driven	SLT-RLT	1200	1800	0.67
Opstal, e.a. 1996	Maasvlakte NL, pile 1	steel HP heavy point, open end	unknown	5350	5800	0.92
Opstal, e.a. 1996	Maasvlakte NL, pile 2	steel HP heavy point, closed end	unknown	6500	7100	0.92
Opstal, e.a. 1996	Maasvlakte NL, pile 3	steel HP heavy point, closed end	unknown	6400	7500	0.85
Opstal, e.a. 1996	Maasvlakte NL, pile 6	concrete with casing	unknown	4400	5000	0.88
Opstal, e.a. 1996	Maasvlakte NL, pile 8	concrete with casing	unknown	4630	4800	0.96
Opstal, e.a. 1996	Maasvlakte NL, pile 10	concrete with casing	unknown	4290	5100	0.84
Holeyman e.a. 2003	Limelette B	de Waal	nearby pile	2200	2850	0.77
Holeyman e.a. 2003	Limelette B	Fundex	nearby pile	2850	3000	0.95
Holeyman e.a. 2003	Limelette B	Omega	nearby pile	2700	2550	1.06

Table 9 Overview results drilled piles in sand.

Source	Location	Pile type	Test order	Static test kN	Rapid test kN	Model factor
Briaud 2000	sand, pile 2	bored pile with defects	RLT-(DLT)-SLT	1602	2460	0.65
Presten 2001	mixed, pile site 1 middle	drilled pile	RLT-SLT	5200	3300	1.58
Middendorp 2008	Waddinxveen pile 48	CFA	RLT-SLT	920	742	1.24
Middendorp 2008	Waddinxveen pile 49	CFA	RLT-SLT	840	1221	0.69
Middendorp 2008	Waddinxveen pile 50	CFA	RLT-SLT	820	649	1.26
Middendorp 2008	Waddinxveen pile 51	CFA	RLT-SLT	540	503	1.07
Middendorp 2008	Waddinxveen pile 52	CFA	RLT-SLT	840	754	1.11
Middendorp 2008	Waddinxveen pile 53	CFA	RLT-SLT	840	653	1.29

Chapter 10

Draft standard for execution of a rapid load test

1 SCOPE

This standard establishes the specifications for the execution of rapid pile load tests in which a single pile is subject to an axial load in compression of intermediate duration to measure its load-displacement behaviour under rapid loading and an assessment of its static behaviour.

The provisions of this standard apply to piles loaded axially in compression.

This standard provides specifications for:

1 Investigation tests, whereby a sacrificial pile is loaded up to ULS (Ultimate Limit State).
2 Control tests, whereby the pile is loaded up to a specified load in excess of the SLS (Serviceability Limit State).

The magnitude of the rapid load tests must make allowance for the influence of rapid phenomena, such as rate effects, pore water pressures and acceleration, when comparing with the equivalent static load test.

Notes:

– generally, an investigation test focuses on general knowledge of a pile type; a control test focuses on one specific application of a pile
– While selecting the load in the rapid load test or during interpretation, one needs to take account of the effects in which the rapid load test differs from an equivalent static load test, such as rate effects, pore water pressures, creep and acceleration.

2 NORMATIVE REFERENCE

The following referenced documents are indispensable for the application of this Standard:

EN 1990: 2002, Euro code 0: Basis of structural design
EN 1997-1, Euro code 7: Geotechnical design – part 1: general rules
NEN-EN-ISO 22477-1 Geotechnical investigation and testing – Testing of geotechnical structures – part 1: Pile load test by static axially loaded compression.

3 TERMS, DEFINITIONS AND SYMBOLS

3.1 Types of piles

Test pile: a pile to which loads are applied to determine the resistance and/or the deformation characteristics of the pile and the surrounding soil.

- trial pile: a pile installed to asses the practicability and suitability of the construction method
- preliminary pile: a pile installed before the commencement of the main piling works or a specific part of the works for the purpose of establishing the suitability of the chosen type of pile and for confirming its design, dimensions and bearing resistance
- working pile: a pile for the foundation of the structure

A test pile can be a trial pile, a preliminary pile or a working pile.

3.2 Rapid load

The load on the pile can be considered as a rapid load if the duration of the load fulfils the following

$$10 \prec \frac{T_f}{L/c_p} \leq 1000 \tag{1}$$

The force shall be applied in a continuously increasing and continuously decreasing manner and should be sufficiently smooth.

3.3 Equivalent diameter

The equivalent diameter of a pile is

- for a circular pile the outer diameter of the pile
- for a square pile the diameter which gives the same area as the square pile (as long as the longest side is smaller than 1.5 times the shortest side)
- for other piles the diameter which is used in static calculations of the pile

3.4 Reference distance

The reference distance is the distance which separates a stationary reference point from a point that will be significantly displaced by the testing method. Only stationary points can be used for reference of displacement measurements. Displacement measuring systems may be placed on the soil outside the reference distance without isolating (displacement compensating) measures.

The value of the reference distance is the maximum of

- measured from the pile: the distance which the waves in the soil travel during the loading (T_f);
- measured from equipment with a falling mass: the distance which the waves in the soil travel during the falling of the mass and the subsequent loading (T_f).

3.5   Failure of a pile

Failure of a pile refers to geotechnical failure according EC 7 part 1.

3.6   Symbols

a acceleration [m/s^2]
c velocity of a stress wave [m/s]
c_p velocity of the stress wave in the test pile [m/s]
c_s velocity of the shear wave in the soil [m/s]
c_R velocity of the surface wave in the soil [m/s]
D (equivalent) diameter of the test pile [m]
e eccentricity [m]
f frequency [Hz]
F force [N]
g acceleration due to gravity [9.81 m/s^2]
L length of the test pile [m]
r distance [m]
r_{ref} reference distance [m]
T_f duration of the load [s]
u displacement [m]
v particle velocity [m/s]
   Prefixes accepted by SI-units (such as [mm] in stead of [m] and [kN] in stead of [N]) are accepted.

4   EQUIPMENT

4.1   General

This standard is applicable to all types of equipment able to generate a loading at the pile head which fulfils the requirements in Section 3.2. If information on the ultimate bearing capacity of the pile is one of the goals of the test, the equipment must have enough capacity to reach the bearing capacity in rapid loading. The annex gives information on equipment which may fulfil the requirements.

Note:
The required force on the pile head during a rapid load test for measuring the ultimate bearing capacity may exceed that for static loading by a factor of 2 in certain soil types.

If for a Rapid Load Test, one or more of the requirements mentioned in this Standard is not met, it should be proven that this shortcoming has no influence on the achievement of the objectives of the test, before the results can be interpreted as a rapid load test.

4.2   Loading

The selection of the loading equipment shall take into account

- the aim of the test
- the ground conditions
- the maximum pile load
- the strength of the pile (material)
- the execution of the test
- safety aspects

The loading equipment must be able to generate a force which fulfils the requirements in 3.2 and generates the required maximum force.

If a test pile is tested by several cycles, the maximum force of each cycle should aim to be larger than the maximum force of the preceding cycle.

The equipment must be able to load the pile accurately along the direction of the pile axis. The eccentricity of the load must be smaller than 10% of the equivalent diameter. The deviation of the alignment of the force to the axis of the pile must be smaller than 20 mm/m

The stress in the pile under the maximum applied load must not exceed the permissible stress of the pile material.

Rebound of the mass on the pile head is not allowed without measurement of the resulting pile head load, deflection and acceleration.

4.3   Measurements

During a rapid load test three variables must be measured:

- the force applied to the pile head
- the displacement of the pile head
- the acceleration of the pile head

The transducers and signal processing must satisfy the following requirements.

General	Requirement
sample frequency	> 4 kHz
duration of pre-event	> 50 ms
duration of post-event	> 300 ms
cut off frequency low pass filter	> 1 kHz

(Continued)

Load

max. load	> target load
linearity	< 2 % of maximum value reached
hysteresis	< 2 % of maximum value reached
response time	< 0.1 ms

Acceleration

number of transducers	≥ 1
resonant frequency	> 5 kHz
linearity	up to 50 g

Displacement

range	> 50 mm and D/20
accuracy	± 0.25 mm
response time	< 0.1 ms
reference distance	> 15 m and c_s/T_f

Before and after each cycle, the level of the pile head must be determined relative to a point outside of the reference distance by high precision optical levelling.

The base of a displacement measuring system should not be placed within the reference distance (see Section 3.4) from the pile. This must be verified at the test site. If the reference distance for a displacement measuring system cannot be reached, the displacement measuring system must be placed on a vibration free base.

The velocity of the pile head shall be calculated by integration of the measured accelerations with respect to time. Calculation of the displacement of the pile head by double integration of the measured accelerations with respect to time is allowed only if the final set is checked by a direct measurement of the displacement.

All equipment used for measuring load, displacement and acceleration in the test must be calibrated. The equipment must be checked on a regular basis. The results of these checks must be registered and kept with the most recent calibration. This data must be made available prior to commencement of the test.

Note:

The time between the checks and calibrations is not prescribed, since the duration of validity of a calibration may depend strongly on the type of measurement device. However, the checks must be of that detail that one can be sure that all measurement devices are showing valid values during the test. It is preferred that all checks are carried out directly before the test, to avoid influence of transport and time.

All loadings (larger than 1 % of the expected static bearing capacity of the pile) after installation of the pile must be measured. This includes all types of static preloading of the pile.

The measurement of the load on the pile head by strain gauges which are fixed to the pile, is allowed for internal forces only. The strain gauges must be calibrated against the results of an external load cell.

5 TEST PROCEDURE

5.1 Preparation

In advance of the test, an execution plan shall be formulated. The plan shall include the following:

1   Test objectives
2   Ground conditions, based on soil investigation in accordance with the regulations in EC 7
3   Testing date
4   Locations, types and specifications of the test piles
5   Allowable values of the load on the pile and the pile displacement
6   Required displacement of the pile and load on the pile
7   Specification of the loading device
8   Specifications of the measurement devices
9   Specifications of additional measurement-devices
10 Number of loading steps and target maximum load per step
11 Check on the acceptability of the foreseen load on the pile and displacement of the pile (with respect to the allowable values, defined item 5)
12 Duration of the measurement and sampling frequency
13 Plan of the test site
14 Time schedule
15 Personnel, showing who is responsible for

    a   supervising
    b   safety
    c   loading
    d   data recording
    e   other tasks

16 Safety requirements
17 Legally required licences for handling the equipment
18 Other points of attention

5.2 Safety requirements

5.2.1 For people and equipment in the surrounding

Safety of human beings and equipment in the surrounding area must be assured during execution of the test. The distance between the nearest person and the test equipment must be at least twice the height of the test equipment measured from the soil surface.

People in neighbouring buildings must be informed about testing. Hindrance to vibration sensitive processes in neighbouring buildings shall be prevented.

5.2.2 For the test pile

The test pile should not be damaged by the test. During a rapid load test, the test pile will be loaded with a force which may exceed the static equivalent test loads by a factor of 2. Test piles should be designed to withstand the resulting higher stresses.

For working piles the maximum displacement of the pile head shall be agreed before commencement of the test. It may not exceed 10% of the (equivalent) diameter.

5.3   Preparation of the pile

The pile head must be flat, plane, perpendicular to the pile axis and undamaged.

The integrity and capacity of the pile must be sufficient to carry the planned test load. If installation of the pile causes doubts about pile integrity, the pile should be tested acoustically, or the rapid load test must be carried out by multiple steps with increasing pile load.

Note:
Doubts about the integrity of a pile might be due to unexpected behaviour during construction. One might think of driving resistance, amount of concrete used, progress during drilling. The deviation might be a deviation from expected values or a deviation of a specific pile from other piles constructed at the construction site or similar site.

The test pile must have enough length above the ground surface to attach the measurement devices. All acceleration transducers must be installed firmly against the pile head.

Between the installation of the test pile and the beginning of the test, adequate time shall be allowed to ensure that the required strength of the pile material is achieved and the ground has sufficient time to recover from the process of pile installation and dissipation of pore-water pressures and other aspects, such as mechanical heat from boring or hardening concrete. The required waiting period may be assessed by measurements of e.g. excess pore-water pressure and soil strength evaluation. During this period, the pile may not be disturbed by load, impact or vibration, or other external influence.

The following time periods between installation and testing of a pile are prescribed:

– for trial and preliminary piles: minimum 7 days in non-cohesive soils, minimum 3 weeks for bored piles in cohesive soils and 5 weeks for driven piles in cohesive soils
– for working piles: minimum 5 days in non-cohesive soils, minimum 2 weeks for bored piles in cohesive soils and 3 weeks for driven piles in cohesive soils.

Note:
These time periods are acc. ISO 22477-1 Sec 5.2.3 (draft).

5.4   General preparations

The sensitive parts of the test equipment shall be protected from weather (rain, wind, direct sunlight) and other disturbances.

All components of the system shall be protected against damage during all stages of construction and testing. Special attention must be paid to cables.

Any other site activities that might influence the measurements, e.g. vibrations by nearby traffic or ongoing pile driving, shall be avoided.

5.5   Aftercare of a working pile

If the result of the test causes doubts about pile integrity afterwards, the pile shall be tested acoustically.

Note:
Doubts about the integrity of a pile might be due to unexpected behaviour during rapid load testing. The deviation might be a deviation from expected values or a deviation of a specific pile from other piles tested at the construction site or similar site.

6   TEST RESULTS

The test results shall consist of

– the force of the loading system [N] at the pile head as a function of time [s]
– the displacement of the pile head [m] as a function of time [s]
– the acceleration of the pile head [m/s^2] as a function of time [s]

There shall be a common base to all time measurements.

All test results must be available in charts and digitally in an ascii-format. All results must be corrected for calibration factors. SI units are prescribed. Corrections applied to the measured signals must be put down in writing.

The measurements of pile levels before and after each cycle are reported. All other readings, such as temperature, tests on concrete samples, level readings, pile shape, static tests on the site, when relevant, must be put down in writing.

The rapid load-settlement diagram must be drawn. This diagram shows the measured pile head displacement [m] against the measured pile head force [N], without any correction.

A copy of all results shall be stored on a back-up medium.

7   TEST REPORT

The load test report shall at least comply with EN 1997-1. It should at least include the following information and data:

1   Reference to all relevant standards
2   General information concerning the test site and the test program:

    a   topographic location of the test
    b   description of the site
    c   purpose of the test
    d   test date
    e   the intended and realized testing program
    f   reference to the organization which carried out the test
    g   reference to the organization which supervised the test

3   Information concerning the ground conditions

    a   reference to the site investigation report
    b   location and reference number of the relevant soil tests
    c   description of the ground conditions, in particular at the vicinity of the test pile

4   Specifications concerning the test pile

    a   the pile type, its nomination and its reference number
    b   the topographic location of the test pile
    c   pile data, such as geometry, top and base level, pile material and reinforcement
    d   date of installation
    e   description of the pile installation and any observations related to the execution, likely to have an influence on the test results
    f   installation records, such as driving logs, concrete consumption, drilling progress

5   Specifications concerning the test

    a   the postulated maximum test load
    b   pile cap details
    c   details of the loading apparatus and measuring devices, including the calibration data
    d   the number of load cycles and the foreseen loading levels
    e   information on the potential energy for each cycle (drop height, mass, amount of fuel)
    f   the distance between the pile and the displacement measurement device
    g   details on the installation of the equipment by drawings and/or photographs

6   The test results

    a   as defined in chapter 6, including the digital data
    b   the rapid load-displacement diagram for each cycle from the measured signals
    c   the net settlement of each cycle
    d   the results of the high precision optical levelling

    If the report includes an interpretation of the results with respect to the purpose of the test, the following information must be added

    a   the method used for the interpretation (with reference to the description)
    b   the derived static load-displacement diagram
    c   the treatment of rate effects
    d   the treatment of mass effects
    e   the treatment of effects due to pore water overpressure

ANNEXES

Annex A

Examples of equipment to which the standard is applicable
Informative

A.1   Statnamic

The Statnamic Load Test (STN) has been developed by Profound and Bermingham-mer Foundation Equipment. The principle of the test is based on the launching of a reaction mass by burning fuel in a closed pressure chamber. This reaction mass is only 5% of the weight needed for a static load test. Loading is perfectly axial.

Figure A1-1 represents the successive stages of a Statnamic load test. Phase I is the situation just before launching. A cylinder with pressure chamber has been connected to the pile head and the reaction mass has been placed over the piston. In phase II the solid fuel propellant is ignited inside the pressure chamber, generating high-pressure

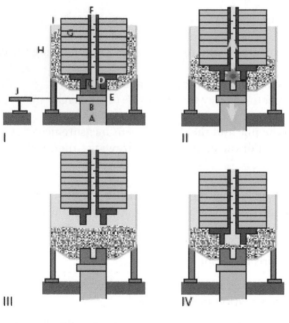

Four stages of a Statnamic test with gravel catch system.

A = pile to be tested   F = silencer
B = load cell   G = reaction mass
C = cylinder & pressure chamber   H = gravel container
D = piston   I = gravel chamber
E = platform   J = optical measuring system

Figure A1-1 Stages of an Statnamic test.

gases and accelerating the reaction mass. At this stage the actual loading of the pile takes place, as an equal and opposite reaction force gently loads the pile. The applied pile force, displacement and acceleration are directly monitored. The upward movement of the reaction mass results in space, which is filled by the gravel (phase III). Gravity causes the gravel to flow over the pile head as a layer, catching the reaction mass and transferring impact forces to the subsoil (phase IV).

Available device loads are 1, 2, 3, 4, 5, 8, 16, 20, 30, and 40 MN. The testing range is between 25% and 100% of the device load.

During the test the reaction mass reaches a height between 2–3 m and then falls back. For high loads of 5–30 MN, gravel is used to catch the reaction mass. For loads in the range of 1–8 MN, a hydraulic catching system is utilized to arrest the reaction mass. With the latter system a considerable shorter cycle time is achieved, enabling more tests per day. With a 4 MN hydraulic catching device, 3–4 piles can be tested per day.

Reference

Middendorp, P., Bermingham, P., Kuiper, B. Statnamic load testing of foundation piles. In: "Proc. 4th Int. Conf. Appl. Stress-Wave Theory to Piles, The Hague, Sept. 1992", Rotterdam, Balkema, 1992, pp. 581–588.

A.2   Pseudo Static Pile Load Tester

The load test by the Pseudo Static Pile Load Tester (PSPLT) is carried out by means of dropping a heavy mass (25.000 kg) with a coiled spring assembly from a predetermined height onto a single pile. After the hit, the mass bounces and is caught in its highest position. Catching the bouncing mass makes larger drop heights possible and avoids further hindrance to the test and the measurements.

The instrumentation for the test consists of a load cell and an optical displacement measuring device. The load cell is placed on top of the pile. It is almost identical to the one used during static load tests. Pile head displacement is recorded with the optical device mounted on a tripod at a distance of approx. 10 m from the pile. This tripod is equipped with a geophone to monitor vibrations of the tripod during the test. All measured signals are immediately processed by a computer and presented in relevant graphs.

The mass effects of the coiled springs in the PSPLT are minimized by using additional rubber springs and by creating a time delay between subsequent coils hitting the base plate. The spring stiffness is order 8 MN/m, but in fact a non-linear spring was installed.

The execution of a test is as follows: the PSPLT is brought to the test site by a low-loader. It moves on its tracks to the test pile, whose pile head has previously been prepared. When the rig is positioned and the measuring devices are attached the test starts. First a static load test is carried out with the weight of the drop mass. Then subsequently a number of rapid loads are deployed to the pile by dropping the mass from increasing heights onto the pile. With the output of results a quasi-static load-settlement curve is produced. Then the next pile can be tested. It is possible to load-test a significant number of piles per single working day. With proper preparations on the test site and the pile heads more than 10 piles daily have been tested.

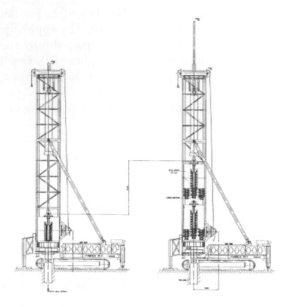

Figure A2-1 Sketch of the PSPLT.

Reference

Schellingerhout, A.J., Revoort, E. Pseudo static pile load tester. In: "Proc. 5th Int. Conf. Appl. Stress-Wave Theory to Piles, Orlando, Sept. 1996", Gainesville, Univ. Florida, Dep. Civ. Eng., 1996, pp. 1031–1037.

A.3   Spring Hammer Test Device

The loading mechanism of the Spring Hammer Test Device (SH device) is similar to that of the Pseudo-Static Pile Load Tester, except that the spring unit is placed on the pile head in the SH test. Two types of the SH device, portable and machine-mounted types, are available as shown in figure A3-1 and Table A3-1. Coned disk springs are used to constitute a spring unit. The performance of the SH device can be easily controlled by changing the combination of the hammer mass and the spring value as well as the falling height of the hammer. One of advantages of the SH device is that repetitive loading can be easily done.

The applied force and the accelerations at the pile head are measured. Direct measurement of the displacement is possible by means of laser displacement meter or optical displacement meter. All the dynamic signals are recorded through a computerised signal acquisition system, and processed to estimate 'static' response of the test pile.

The SH device may be used very effectively to obtain the performance of piles having relatively low bearing capacity.

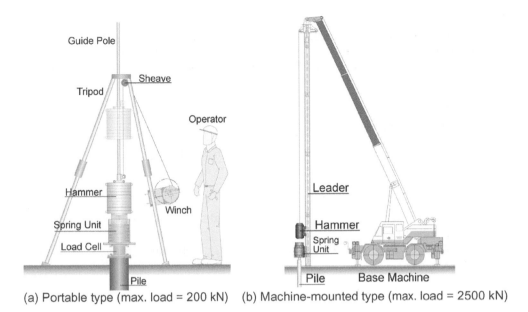

(a) Portable type (max. load = 200 kN)   (b) Machine-mounted type (max. load = 2500 kN)

Figure A3-1 Spring hammer test device.

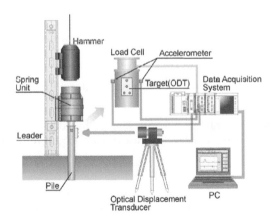

Figure A3-2 Signal acquisition system.

Table A3-1 Standard specifications of spring hammer devices (as in 2007).

	Portable	Machine mounted
Hammer mass (ton)	0.2	3
Spring values (kN/m)	5125	35000
Max. fall height (m)	2	3
Max. load (kN)	200	2500
Weight of spring unit (kN)	1	20
Number of tests per day in usual test condition	8 to 10	5 to 7

Reference

Matsumoto, T., Wakisaka, T., Wang, F.W., Takeda, K. and Yabuuchi, N. Development of a rapid pile load test method using a falling mass attached with spring and damper, In: Proc. 7th Int. Conf. on the Appl. of Stress-Wave Theory to Piles, Selangor, Malaysia: 351–358.

Annex B

Interpretation of the test Informative

The measured force is not equal to the force which will be measured by at SLT. The test results must be interpreted

1   by comparing RLT and SLT at the test site
2   by using empirical relation, based on rapid load tests and static tests at sufficient comparable sites
3   an analytical or numerical model, validated at a sufficient number of load tests and static tests.

The following aspects must be evaluated during interpretation:

–   inertia effect
–   rate effect
–   generation of pore water pressure
–   plug-motion for open ended piles.

For the interpretation reference is made to a number of international documents.
An international guideline written by CUR, BRE, LCPC and WTCB. (to be published in 2009).
Additional information on the interpretation of the test can be found in

with respect to the inertia effect:
Middendorp, P., Bermingham, P., Kuiper, B. Statnamic load testing of foundation piles. In: "Proc. 4th Int. Conf. Appl. Stress-Wave Theory to Piles, The Hague, Sept. 1992", Rotterdam, Balkema, 1992, pp. 581–588.

with respect to rate effects in clay:
Brown M.J., Anderson W.F., Hyde A.F., Statnamic testing of model piles in a clay calibration chamber. In: Int. Jnl. Phys. Modelling Geotechnics., Vol. 4, No. 1 pp. 11–24 (ISSN 1346-213X).

with respect to the generation and influence of porewater pressure in sand:
Huy, N.Q., van Tol, A.F., .Hölscher, P. Interpretation of rapid pile load tests in sand in regard of rate effect and excess pore pressure to be published in Proceedings of the 8th International Conference of the Application of Stress-wave Theory to piles, Lisbon (PT), 8–10 September 2008.

with respect to the plug effects of open ended piles:
Ochiai Ochiai, H., Kusakabe, O., Sumi, K., Matsumoto, T. and Nishimura, S. Dynamic and Statnamic load tests on offshore steel pipe piles with regard to failure mechanisms of pile-soil interfaces at external and internal shafts. Proc. Int. Conf. on Foundation Failures, Singapore, 327–338, 1997.5.12–13.

Annex C

Information on requirements for the load Informative

This annex shows an indicative method to judge the applicability of the applied force, which is defined in Section 3.2. The method is shown graphically in figure C-1. The requirements are defined in terms of the time derivative of the force.

Three times are defined:

t_{start} start of the loading
t_{zero} the time the maximum load F_{max} is reached
t_{end} the end of the loading

The measured force is differentiated with respect to time. The resulting curve can be approximated by two half sine functions, one for the increasing part of the loading ($t_{start} < t < t_{zero}$) and one for the unloading part ($t_{zero} < t < t_{end}$). If the derivative does deviate less than 40% of the amplitude of the sine from the sine function, the loading can be accepted.

For loading equipment which has a different static force before and after the test (e.g. Statnamic starts with a non-zero load, a bouncing system might end with a non-zero load), the linear approximation must be corrected by a linear curve between the two static loadings at t_{start} and t_{end}.

Note:
This method is a proposal to avoid discussion in future. Both the method and the limiting values must be tested against real measured data. Examples will be added.

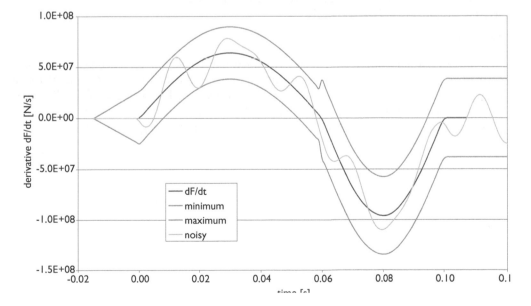

Figure C-1 Calculated derivative of force.

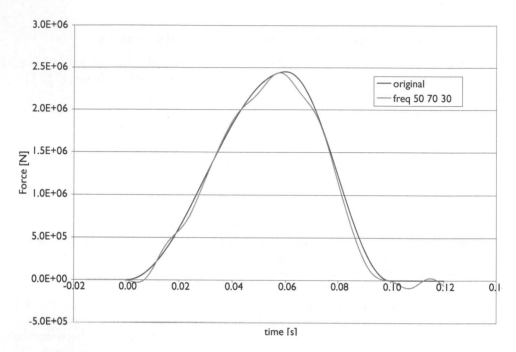

Figure C-2 Force related to the derivative in Figure D-1.

Annex D

Information on requirements transducers and calibration Informative

Accelerometers

The application of servo electrical and piezo-electrical accelerometers is allowed. However, the integration of piezo electrical transducers with respect to time is more complicated. Therefore, it is advised to use servo-electrical transducers if the displacements must be calculated from the measured accelerations.

Moreover, the sample frequency must be higher if time integration is required.